# CAMBRIDGE LIBRARY COLLECTION

*Books of enduring scholarly value*

## Technology

The focus of this series is engineering, broadly construed. It covers techno-logical innovation from a range of periods and cultures, but centres on the technological achievements of the industrial era in the West, particularly in the nineteenth century, as understood by their contemporaries. Infra-structure is one major focus, covering the building of railways and canals, bridges and tunnels, land drainage, the laying of submarine cables, and the construction of docks and lighthouses. Other key topics include developments in industrial and manufacturing fields such as mining technology, the production of iron and steel, the use of steam power, and chemical processes such as photography and textile dyes.

## Rivers and Canals

Leveson Francis Vernon-Harcourt (1839–1907) drew on a distinguished career in canal and river engineering for this illustrated two-volume survey, here reissued in its enlarged 1896 second edition. Having started as an assistant to the civil engineer John Hawkshaw, Vernon-Harcourt was appointed resident engineer in 1866 for new works on London's East and West India docks. Later, as a consulting engineer, he specialised in the design and construction of harbours, docks, canals and river works, and he was elected professor of civil engineering at University College London in 1882. This publication covers the design and construction of tidal and flood defences, canals, locks, and irrigation works. Volume 2 covers canal engineering, discussing the design and construction of canals and their associated works such as locks and lifts. Vernon-Harcourt also discusses ship canals and irrigation works. His *Harbours and Docks* (1885) is also reissued in this series.

Cambridge University Press has long been a pioneer in the reissuing of out-of-print titles from its own backlist, producing digital reprints of books that are still sought after by scholars and students but could not be reprinted economically using traditional technology. The Cambridge Library Collection extends this activity to a wider range of books which are still of importance to researchers and professionals, either for the source material they contain, or as landmarks in the history of their academic discipline.

Drawing from the world-renowned collections in the Cambridge University Library and other partner libraries, and guided by the advice of experts in each subject area, Cambridge University Press is using state-of-the-art scanning machines in its own Printing House to capture the content of each book selected for inclusion. The files are processed to give a consistently clear, crisp image, and the books finished to the high quality standard for which the Press is recognised around the world. The latest print-on-demand technology ensures that the books will remain available indefinitely, and that orders for single or multiple copies can quickly be supplied.

The Cambridge Library Collection brings back to life books of enduring scholarly value (including out-of-copyright works originally issued by other publishers) across a wide range of disciplines in the humanities and social sciences and in science and technology.

# Rivers and Canals

*With Statistics of the Traffic
on Inland Waterways*

VOLUME 2: CANALS

LEVESON FRANCIS VERNON-HARCOURT

CAMBRIDGE
UNIVERSITY PRESS

# CAMBRIDGE
## UNIVERSITY PRESS

University Printing House, Cambridge, CB2 8BS, United Kingdom

Cambridge University Press is part of the University of Cambridge.

It furthers the University's mission by disseminating knowledge in the pursuit of
education, learning and research at the highest international levels of excellence.

www.cambridge.org
Information on this title: www.cambridge.org/9781108080606

© in this compilation Cambridge University Press 2015

This edition first published 1896
This digitally printed version 2015

ISBN 978-1-108-08060-6 Paperback

# RIVERS AND CANALS

*VERNON-HARCOURT*

VOL. II.

𝕷𝖔𝖓𝖉𝖔𝖓

HENRY FROWDE

OXFORD UNIVERSITY PRESS WAREHOUSE
AMEN CORNER, E.C.

𝕹𝖊𝖜 𝖄𝖔𝖗𝖐

MACMILLAN & CO., 66 FIFTH AVENUE

# RIVERS AND CANALS

THE
FLOW, CONTROL, AND IMPROVEMENT OF RIVERS

AND THE

DESIGN, CONSTRUCTION, AND DEVELOPMENT OF CANALS
BOTH FOR NAVIGATION AND IRRIGATION

*WITH STATISTICS OF THE TRAFFIC
ON INLAND WATERWAYS*

BY

LEVESON FRANCIS VERNON-HARCOURT, M.A.

MEMBER OF THE INSTITUTION OF CIVIL ENGINEERS
AUTHOR OF 'HARBOURS AND DOCKS,' AND 'ACHIEVEMENTS IN ENGINEERING'

IN TWO VOLUMES.—VOL. II
CANALS
SECOND EDITION, RE-WRITTEN AND ENLARGED

Oxford
AT THE CLARENDON PRESS
1896

𝔒𝔵𝔣𝔬𝔯𝔡

PRINTED AT THE CLARENDON PRESS

BY HORACE HART, PRINTER TO THE UNIVERSITY

# CONTENTS OF VOL. II

## CHAPTER XV.

### CANAL WORKS.

## CHAPTER XVI.

### CANAL LOCKS.

## CHAPTER XXI.

### INLAND NAVIGATION (*continued*).

## CHAPTER XXII.

### FORMS OF BARGES; METHODS OF TRACTION; AND SHIP-CANALS
### INSTEAD OF RIVERS.

## CHAPTER XXIII.

### SHIP-CANALS FOR PORTS.

# CHAPTER XXIV.

## INTEROCEANIC SHIP-CANALS.

# CHAPTER XXV.

## INTEROCEANIC SHIP-CANALS (*continued*).

# LIST OF ILLUSTRATIONS IN VOL. II

—⊷—

## CANALS.

### PLATE XI.

#### CANAL WORKS AND SECTIONS.

##### CANAL WORKS.

Locks on the St. Denis Canal, France. Plan and Sections.—Lock-Gates.
Elevation and Section.
Lock on the Scheldt and Meuse Canal. Plan and Sections.
Georgetown Incline, United States. Elevation and Section.
La Louvière Lift, Belgium.

##### CANAL SECTIONS.

Inland Canals. Monkland.—Wilts and Berks.—Grand Junction.—Glasgow
and Paisley.—Forth and Clyde.—Marne and Rhine.
Ship-Canals. Gloucester and Berkeley.—Caledonian,—Manchester.—Tancarville.—Terneuzen.—North Holland.—Amsterdam, through Sand Hills and
through Lakes, original and enlarged.—Baltic, in High, and Low Ground.
—Panama, in Earth, and in Rock.—Corinth.—Nicaragua, in Earth, and
in Rock.—Suez, original, and enlarged.

### PLATE XII.

#### SHIP-CANALS FOR PORTS.

Amsterdam Ship-Canal. Plan and Longitudinal Section.—Ymuiden and
Zuider Zee Locks. Plans.—Ymuiden Breakwaters. Section.—Zuider Zee
Embankment. Section.
Manchester Ship-Canal. Plan and Longitudinal Section.—Eastham Locks.
Plan.—Latchford Locks and Sluices. . Plan and Section.
Baltic Canal. Plan and Longitudinal Section.—Elbe Locks. Plan.—Kiel
Locks. Plan.

## PLATE XIII.

### ISTHMIAN SHIP-CANALS.

**Suez Canal.** Plan and Longitudinal Section.—Port Said Breakwater. Section.
—Port Said Harbour. Plan.
**Panama Canal.** Plan and Longitudinal Section, original, and with Locks.
**Corinth Canal.** Plan and Longitudinal Section.
**Nicaragua Canal.** Plan and Longitudinal Section.
**Central America.** Routes of Canals and Ship-Railway.

---

## ERRATA.

Page 349, line 16 from top, *for* Gironde *read* Garonne.
   „       „  .17 from bottom, *for* Gironde into the Garonne *read* Garonne into
                    the Gironde.
  „ 363 „   15 from top, *for* Burgogne *read* Bourgogne.
  „ 366 „   5 from bottom, *for* Burgogne *read* Bourgogne.
  „ 550 „   14 from top, *for* two *read* three.
  „ 621 „   4 from bottom, *for* Azov *read* Azof.
  „ 622 „   3 from top, *for* Azov *read* Azof.
  „ 625 „   1 from bottom, *for* Columbian *read* Colombian.
  „ 630 „   16 from bottom, *for* Four *read* Five.

# RIVERS AND CANALS.

## PART II. CANALS.

## CHAPTER XV.

### CANAL WORKS.

Objects of Canals. Drainage Canals: Purposes and Form. Irrigation Canals: Object, Sources of Supply, Value, Instances. Navigation Canals: Objects, Instances. Canals connecting Rivers: Various Examples. Lateral Canals: Uses, Instances, in place of Tidal Navigations, when preferable to Canalizing a River. Sizes of Canals for Navigation: Dimensions of various Inland Canals and Ship-Canals; Minimum Section in relation to Barges; Average Dimensions of Barges on British Inland Canals, and of Vessels on Main Lines of French Canals. Construction of Canals for Navigation: compared with Railways; Watertight Trench; Earthwork for Canals, Dredgers and Excavators for large Excavations, Protection of Slopes from Slips; Towing-Path. Bridges over Canals: similar to Railway Bridges. Swing-Bridges: Forms, Examples, Advantages of Steel in construction. Bascule Bridges: Description, Examples in Holland, Tower Bridge. Lift-Bridges: Method of Working, Examples. Aqueducts for Canals: Instances, Barton Swing Aqueduct. Tunnels for Canals: Examples in England and France. Waste Weirs, and Sluices. Stop-Gates. Supply of Water. Reservoirs: Instances in Great Britain, France, and Russia. Reservoir Dams: of Earth, of Masonry; Construction, and Auxiliary Works. Introduction of Supply of Water. Consumption of Water in Canals. Remarks on Canal Works.

CANALS are artificial channels containing water; and they are constructed either for the conveyance of water, or to provide inland waterways for navigation. In the first case, the channels are given a definite fall, so as to ensure the flow of the water along them; and they either serve to improve the drainage of low-lying lands, when the natural watercourses are insufficient and the available fall is small, or are employed for conveying water from rivers for irrigating lands in hot countries. The channels, on the contrary, which are formed to provide inland waterways, are made level, in order to retain the water for navigation.

**Drainage Canals.** Reference has been previously made to the construction of numerous straight drains in the flat, low-lying Fens, for preventing floods and improving the drainage of these districts (pp. 154 and 284). These drainage canals are in reality artificial rivers, made straight in order to utilize the small fall as much as possible, and serving, like natural watercourses, to discharge the rainfall from the adjacent lands. Their sectional area is made proportionate to the discharge they are required to provide with the available fall, as determined by the discharge formulæ given in Chapter II, which are essential for the correct design of such channels. Drainage canals, being restricted to low-lying lands with a small fall, are not generally traversed by currents of sufficient velocity to erode the bed or sides of the channel. These canals are sometimes utilized for navigation; but it is difficult to satisfy fully the somewhat conflicting interests of drainage and navigation above the tidal limit, as it is expedient for efficient drainage to lower the water as much as possible, whilst for the sake of navigation the water should be maintained at a certain level. In tidal channels, however, the interests of drainage and navigation are both benefited by the formation of a direct, deepened channel to the sea (pp. 286–289).

**Irrigation Canals.** In many hot countries, where the rainfall is deficient, and where sometimes no rain falls for long periods, the fertility of the soil and the cultivation of certain crops are dependent on artificial irrigation. Irrigation canals generally draw off a portion of the discharge of a river, and convey the water to the lands to be irrigated, where it is distributed over the land by numerous small branch canals controlled by sluices; and movable dams are provided at the head of the irrigation canal, for arresting

the influx from the river when the water in the river
rises above a certain height. Sometimes, when the water
in the river is low, the flow into the irrigation canals is
facilitated by raising the water-level in the river by closing
a dam placed across the river. Occasionally, in mountainous
districts, a reservoir is formed by a dam across a suitable
part of the valley; and the water thus stored up is conveyed by a canal to irrigate parched districts when required.
The inclination of these canals has to be regulated according to the general available fall of the country traversed; and
the sectional area depends on the amount of this fall, and
the supply of water to be provided. The slope, however,
must be adjusted so that the velocity of flow may not be
liable to produce erosion of the bed of the canal; for
neglect of this consideration, or an error in the calculations
for the discharge, in the case of the Ganges Canal, led to
the erosion of the sandy bed of the canal; and weirs had
to be erected at intervals to reduce the velocity of flow.
On the other hand, the rate of flow must not be so feeble
as to occasion deposit of sediment in the canal; and a small
slope necessitates a larger canal for any given discharge.

The actual commercial value of such works depends upon
the dryness of the district, the nature of the soil and of
the suitable crops, and the extent of land irrigated in
relation to the cost of the works. There is, however, no
doubt that irrigation works have greatly increased the productiveness of the soil in hot, dry districts, and have
converted arid plains into fertile lands; and the value of
these works, when carried out by the government for the
general benefit of the community, by increasing the
food supply of the district, as in the case of the rice
crops in India, and thus mitigating famines, cannot be
measured merely by the annual return on the original
outlay. The prosperity of Egypt is wholly dependent upon
the fertility of the zone of land watered by the Nile and

its numerous auxiliary irrigation canals, and covered annually by a layer of the fertilizing mud brought down by the river from the mountains of Abyssinia. Italy has derived great agricultural benefits from irrigation works, the largest recent example of which is the Cavour Canal, irrigating a large tract of land between Turin and Novara. This canal derives its supply from the river Po, which is supplemented by a supply from the Dora Baltea, a river of glacier origin, when the discharge of the Po becomes small in the summer.

The flow of the rivers in Spain falls very low in the summer, so that there is a dearth of water at the time when it is most required for irrigating the land. This has led to the formation of reservoirs for storing up the water of the rivers, so that it may be available for irrigation in the summer.

Irrigation works have, however, been carried out on the largest scale in India, where long periods of drought, and the density of the population render irrigation essential for ensuring adequate crops. The Ganges, Jumna, Agra, Orissa, Sone, Madras, Godavery, Kistna, and Bari Doab canals are some of the most important irrigation works which have conferred such important benefits on India; and most of the larger irrigation canals have been constructed so as to be available for navigation.

Irrigation canals, indeed, resemble conduits for water-supply on a large scale, for they convey the water from the source of supply to the place where it is utilized; and the volume is determined by the amount required per acre, depending upon the climate, the soil, and the crops, and the extent of land proposed to be irrigated. As, however, the water is only required for agricultural purposes, the purity of the water is immaterial; and in fact a certain amount of sediment, if spread over the land, is often beneficial for the crops.

**Navigation Canals.** Generally canals constructed for the purposes of navigation are totally distinct works from drainage and irrigation canals, although these latter are sometimes utilized for navigation. Before the advent of railways numerous inland canals were constructed in France, Italy, Holland, Belgium, Great Britain, North America, China, and elsewhere, for facilitating the conveyance of goods, owing to the great reduction in the tractive force required on water as compared with traction along ordinary roads. These canals were formed in a series of level reaches ; and the differences in elevation of the reaches were surmounted at their extremities by locks or inclined planes. They were made for connecting centres of population where no navigable rivers existed, and for connecting two adjoining river navigations by traversing the lowest point of the ridge separating their valleys ; and lateral canals have also been constructed to provide a waterway in place of a portion of a river, where, owing to rocky shoals or a very rapid fall, the river presents serious obstacles to navigation.

The canal works carried out during the latter half of the eighteenth, and the earlier part of the nineteenth century, established water communication between the principal towns of England and the chief seaports ; and Paris has been connected by water with the principal towns of France and the coalfields of Belgium, by means of canals and river navigations. New York possesses access by water to the large inland lakes by the Hudson River and the Erie Canal, and is connected with the Delaware River by the Morris Canal ; whilst a through route by water has been formed between Lake Michigan and the Gulf of Mexico, by the construction of the Illinois and Michigan Canal, which connects the lake with the Illinois River, and consequently with the Mississippi.

**Canals connecting Rivers.** Canals joining two river basins furnish a valuable means of providing through routes by water, and of promoting traffic for long distances along river navigations. Even in Great Britain, where the river navigations are necessarily small, owing to the restricted area of the river basins, there are some instances of such canals, namely, the Thames and Severn, the Forth and Clyde, the Trent and Mersey, and the Kennet and Avon canals. In France and Germany, where inland navigation has been much more fully extended than in England, owing to the greater distances of parts of these countries from the sea-coast, the larger size of the river basins, and the attention paid by their governments to the development of the waterways, there are numerous examples of the connection of two river navigations by a canal. Thus the Marne-Rhine Canal, the Rhone-Rhine Canal, and the Oder-Spree Canal connect the rivers from which they derive their names; the Canal de l'Est connects the Meuse, or Maas, with the Saône, and consequently with the Rhone; the Loing and Briare canals, join the Seine and the Loire; and the Bourgogne Canal unites the Yonne and the Saône, and, consequently, the navigations of the Seine and the Rhone, and thereby provides a continuous inland waterway stretching from the English Channel to the Mediterranean. The Danube, also, is placed in communication with the Rhine, by the Main-Danube Canal and the canalized Main, so that there is a continuous waterway between the North Sea and the Black Sea. In Russia, the canal between the Neva and the Volga forms a connection by water between the Baltic and the Caspian; whilst the proposed Don and Volga Canal would extend this communication to the Black Sea. In North America, the Hudson River is connected with the Richelieu River, a tributary of the St. Lawrence, by the Champlain Canal and Lake Champlain. The St. Lawrence, moreover, is placed in communication, by water, with the Mississippi and the Gulf of Mexico through the large lakes, the connecting canals and river navigations, and the Illinois and Michigan Canal, providing an unbroken waterway half across North America, and down the entire length of the United States from north to south.

**Lateral Canals.** Canals are formed alongside rivers where an insurmountable obstacle, such as the falls of Niagara, precludes the conveyance of navigation along a portion of a river, or where rapids, a shallow, shingly, shifting bed, or a torrential flow in a river, render the construction of a lateral canal, supplied with water from the adjacent river, easier or more convenient for navigation than the canalization or improvement of the river. The Welland Canal provides an artificial waterway between Lake Erie and Lake Ontario, where the natural connection along the Niagara River is barred by the falls of Niagara and the rapids below.

Numerous examples exist in France where the navigation follows a lateral canal, constructed alongside a portion of a river, especially in the higher parts of the river basins, where the increased fall and a more torrential flow render most rivers less adaptable for navigation. Thus, for instance, the lateral canal of the Gironde extends from Toulouse at the starting point of the old Languedoc Canal, now called the Canal du Midi, down to Castets near the outlet of the Gironde into the Garonne, a distance of 120 miles. A lateral canal has been constructed along the upper part of the torrential Loire from Roanne to Briare, a distance of 154 miles, which constitutes the only waterway of the first class in the valley of the Loire, owing to the rapid current of the river during floods, and the scarcity of water at other times; and this canal, by its connection with the Upper Seine through the Briare and Loing canals, forms one of the important lines of water communication converging upon Paris, and consequently possesses a large traffic. The lateral canal of the torrential Marne extends from Chaumont, on the Upper Marne, down to Dizy, a length of 109 miles; and a canal is in course of construction for prolonging this waterway upwards beyond the head of the Marne basin, and connecting it with the Saône.

These lateral canals are not wholly confined to inland waterways, but have sometimes been constructed in con-

nection with tidal navigations, as for instance the Gloucester
and Berkeley Canal providing a direct and still-water
navigation up to Gloucester, in place of the circuitous route
of the upper portion of the tidal Severn with its rapid
currents. The Tancarville Canal furnishes a sheltered
waterway between the port of Havre and the trained
channel of the tidal Seine, thereby enabling vessels to
avoid the open estuary; and the Loire lateral ship-canal
supplies a direct, deep channel between the lower estuary
and the trained river, in place of the shallow, circuitous
channel of the upper estuary (Plate 9, Figs. 1, 4, and 5).

Brindley, the engineer of many of the earlier English
canals, considered that the only value of rivers was to
supply water to canals; but in reality the choice between
canalizing a river or constructing a lateral canal, depends
upon the comparative cost. Some rivers have such rapid
currents, such an irregular flow, such high floods, and
such a changeable course, that the construction of a lateral
canal along a portion, at any rate, of the valley, may involve
no greater expense, or be even less costly, than a corresponding improvement of the river for navigation. In
such cases, the formation of a lateral canal is preferable,
as the ascending traffic is not impeded by any current,
and a canal is not exposed to floods, which at times put
a stop to river navigation. Sometimes, also, when the
requirements for the traffic are small and the river is wide,
it is possible to construct a canal, just adequate to provide
the necessary accommodation, at a lesser cost than the works
for canalizing the river would entail. The lower portions,
however, of torrential rivers of moderate size are sometimes
canalized, as for instance two of the torrential tributaries
of the Seine, the Marne and the Yonne, which have been
rendered navigable in this manner for 203, and 67 miles
respectively above their confluence with the Seine. The
torrential Lower Rhone, on the contrary, was improved for

navigation by training works, in preference to the formation of a lateral canal as proposed in two or three schemes.

**Sizes of Canals for Navigation.**  The trench excavated for a canal is generally formed with a flat bottom, and sloping sides with a minimum inclination of $1\frac{1}{2}$ to 1, except through rock, and varying with the nature of the soil (Plate 11, Figs. 13 to 32).  The width and depth of a canal are regulated according to the size of the largest barges or vessels for which provision has to be made.  Some of the small inland canals were only given a depth of water of $3\frac{1}{2}$ to 4 feet and a bottom width just sufficient for two barges to pass; though the ordinary dimensions of the main English inland canals are 5 feet depth of water, 25 feet bottom width, and 40 to 45 feet surface width (Plate 11, Figs. 14 and 15).  Within recent years, the main lines of inland canals in France have been gradually reconstructed, where necessary, so as to afford a minimum depth of water of $6\frac{1}{2}$ feet, and a bottom width of 33 feet (Plate 11, Fig. 21).  A few of the more important early inland canals were made of a larger section, such as the Monkland, Glasgow and Paisley, and Forth and Clyde canals, with depths of water of 8 to 10 feet (Plate 11, Figs. 13, 16, and 17); whilst the Gloucester and Berkeley, Caledonian, and North Holland ship-canals, constructed in the early part of the nineteenth century, and the precursors of the large ship-canals of the present day, were given depths of 18 to 20 feet, and surface widths of $86\frac{1}{2}$ to $123\frac{1}{2}$ feet (Plate 11, Figs. 18, 19, and 24).  Some of the more recent ship-canals to ports have been made very similar in depth; but their width has been augmented.  Thus the Terneuzen and Amsterdam ship-canals were given depths of $21\frac{1}{3}$, and 23 feet respectively, and surface widths of 183, and 176 feet; but the Amsterdam Canal is being enlarged (Plate 11, Figs. 23 and 25).

The depths of the most recent large ship-canals have been regulated by the depth adopted for the Suez Canal, which was originally made 26 feet, and is now being deepened to 29½ feet (Plate 11, Fig. 32). Thus the Manchester and Corinth ship-canals have been made 26 feet deep, and the Baltic Ship-Canal is to have a depth of 29½ feet; whilst the depth proposed for the Panama and Nicaragua canals is 28 feet (Plate 11, Figs. 20, and 26 to 31). The Suez Canal was constructed with a bottom width of 72 feet, admitting the passage of vessels in only one direction at a time, so that passing places of greater width had to be provided at intervals to allow of the passage of vessels in opposite directions. A similar bottom width has been provided in the Corinth and Baltic canals, and in the completed portion of the Panama Canal; but the Suez Canal is being widened to a bottom width of about 230 feet; whilst the Amsterdam and Manchester ship-canals were constructed with bottom widths of 88½, and 120 feet respectively, to enable vessels to pass each other at any part of the canal.

The bottom width of a canal should generally not be less than double the width of the widest barge navigating it, so as to enable two barges to cross; and the depth of water should be at least 8 inches to 1 foot in excess of the greatest draught of any barge passing along it. The barges navigating the inland canals of Great Britain have, for the most part, widths comprised between 7 feet and 15 feet, and draughts of between 3 and 5 feet; and their most common length is about 72 feet[1]. The main lines of canals in France have been reconstructed, since 1879, so as to accommodate vessels of 250 to 300 tons, 126⅜ feet long, 16½ feet wide, and drawing 6 feet of water, which it is hoped may be eventually augmented to 6⅔ feet. In carrying canals through towns, vertical masonry side walls are built on each side to economize land and to provide quays; and in tunnels under bridges, and along aqueducts, vertical sides are also adopted, and the canal is often reduced

[1] 'Returns made to the Board of Trade, in respect of the Canals and Navigations in the United Kingdom, for the year 1888.' London, 1890.

in width, so as only to admit the passage of a single barge. The section, however, of the canal in such cases should not be reduced to less than double, or at the utmost one and a half times the cross-section of the barges navigating the canal, to avoid the great increase in resistance to traction occasioned when the cross-section of the barge approximates to that of the contracted waterway. In order that the resistance to traction in a canal may not be materially greater than that experienced in hauling a barge through a large expanse of water, the bottom width of the canal should be double the maximum width of the barge, the depth $1\frac{1}{2}$ feet more than the draught of the barge, and the section of the canal six times the greatest cross-section of the barge.

**Construction of Canals for Navigation.** The works for a canal resemble those which have become so familiar in the construction of railways. Thus canals have to traverse high ground in cuttings, and to be carried across valleys on embankments; bridges and aqueducts have to be constructed for carrying roads or railways over and under them; and occasionally tunnels have to be made for conveying a canal through intervening hills, especially in traversing the water-parting of two adjacent river basins at the summit-level of the canal.

Whereas a railway is laid to a variety of gradients, with level portions only where the lie of the country is suitable, a canal has to be made in a series of level lengths, with abrupt variations in level between them, which are surmounted by locks, inclines, or lifts. Greater care has, accordingly, to be taken in laying out a canal, as it cannot follow so closely the undulations of the land as a railway does; and it is necessary for it to wind along the contour lines of the slopes of a valley, or to keep in the bottom, so as to avoid as far as possible changes of level, except where the natural fall of the land necessitates an alteration in the level of the canal and the introduction of one or more locks.

When a canal is excavated through porous soil, it has to be

made watertight by a lining of clay, concrete, or other suitable material. The cheapest portion of a canal is where the canal is made partially in cutting, with just sufficient excavation for forming suitable side banks. These banks have to be brought up carefully in layers, and to be made watertight by a puddle wall in the centre, from two to three feet thick, going down a little into the solid ground. The whole channel requires most carefully lining when it is placed on an embankment, to avoid leakage; and special care must be taken in forming the embankment, to prevent subsequent settlement. Where the bottom of the valley, along which a lateral canal has to be formed, consists of alluvial plains, it is advantageous to carry the canal along the base of one of the side slopes of the valley, for the soil there is generally less permeable than the porous soil of the plains. Moreover, this position is more suitable for forming level reaches, and is more secure from the inundations to which flat plains bordering a river are commonly subject; and it also renders easier the crossing of tributaries and watercourses flowing down to the main river, which traverse the route of the canal.

**Earthwork for Canals.** Owing to the necessity of constructing canals for navigation in a series of level reaches, and the excavation of a trench for the waterway, the earthwork required for making a canal is considerably greater than for a railway of similar length, especially in the case of ship-canals, where it is expedient to introduce as few locks as possible. Accordingly, machinery has been resorted to as much as practicable in the larger canal works, to facilitate and expedite the excavation and deposit of the material. Thus dredgers have been employed as soon as water can be admitted to a portion of a ship-canal, for enlarging and deepening the trench, as for instance at the Suez Canal and in the Panama Canal works; and excavators have been very extensively used in the large cuttings involved in the construction of recent ship-canals, such as the Panama, Corinth, and Manchester canals (Plate 12, Fig. 8, and Plate 13, Figs. 6 and 8), many new types of excavators having been designed and employed for the Panama Canal works, and about one hundred excavators having been simultaneously at work in forming the Manchester Ship-Canal. Long shoots for transporting the material lifted by the dredgers, and discharging it on to the banks at the side, were first employed on the Suez Canal works, and subsequently at the Panama Canal works; whilst lines of floating tubes were first adopted for a similar purpose at the Amsterdam Canal works, and afterwards in the enlargement of the Ghent-Terneuzen Canal (pp. 93–98).

The slopes of the cuttings and embankments for canals have to be regulated according to the nature of the soil; and they have to be efficiently drained, and their surfaces protected from the weather, to avoid slips, similarly to railway cuttings and embankments. A slip, however, in a canal embankment is attended with more serious consequences than in a railway embankment, owing to the escape of water resulting from any dislocation of the trench; whilst the slopes of a canal at, and below the water-level, are more subject to injury, and more difficult to repair, than the slopes of a railway cutting, and consequently require special care in construction and maintenance. The slopes of a canal, moreover, near the water-level are exposed to the erosive action of waves generated by the passage of vessels, or occasionally by wind blowing along the canal. Berms have, accordingly, been often introduced at the sides in the larger canals, a little below the water-level, to reduce the waves caused by large vessels passing along at a rapid rate, by widening the water surface of the canal, and thus spreading the undulations over the shallow enlargements (Plate 11, Figs. 18, 22, 25, 26, and 32). Pitching, timber-work, or fascines are also often employed for protecting the exposed portions of the slopes of a canal near and below the water-level, more particularly where vessels are allowed to navigate the canal at a fair speed by the aid of steam power.

A towing-path has to be provided for traction by horses along one side of a canal, about two feet above the water-level; and it is advisable to give the path a slight downward inclination away from the canal, so as to afford the horse drawing the barge a better foothold.

**Bridges over Canals.** Bridges carrying roads or railways over canals are generally precisely similar in construction to the bridges built for railways, or for carrying roads across streams and rivers. They are formed of brickwork, masonry, iron, or steel; and their span depends upon the width of the canal and towing-path, and the angle at which the crossing is effected.

When, owing to the position of the canal, it is inconvenient to place the bridge at a sufficient elevation to afford the requisite headway for vessels, especially in the case of ship-canals traversed by vessels with high masts, movable bridges are constructed. Three types of movable bridges have been employed for carrying roads or railways at a low level over canals, namely, swing-bridges, bascule bridges, and lift-bridges.

**Swing-Bridges.** Swing-bridges supported on a central, or nearly central axis, with the shorter tail-end in the latter case counterpoised by the aid of iron kentledge, make a quarter of a revolution on a central pivot and a ring of rollers, or on the water centre of a hydraulic press, and leave the waterway perfectly free for the passage of vessels; and they close again across the canal, by a reverse movement, for the circulation of traffic over the bridge. This type of bridge is easily and rapidly swung by hydraulic machinery; and it has been extensively adopted for crossing canals, rivers, and the passages and locks of docks. When the canal or navigation is wide enough to place the pier on which the bridge revolves in the centre of the channel, the canal can be economically spanned by a symmetrical cantilever swing-bridge, supported in the middle on the central pier, with its two arms stretching across to either bank; and the bridge on revolving through a quadrant, opens the waterway on each side of the pier for the vessels passing up and down.

Swing-bridges carry two railways and a road across the Amsterdam Ship-Canal; but in these bridges, the central pier is placed at the toe of one of the side slopes of the canal; and one arm spans the deep waterway, whilst the other extends across the slope, the slope on the opposite side being crossed by a fixed span resting on an abutment on the bank and a pier at the toe of the slope. Swing-bridges, and a swing aqueduct at Barton, carry six roads and the Bridgewater Canal across the Manchester Ship-Canal, the canal being widened where a central pier is introduced; but two roads and the various railways cross the canal on high-level bridges, the headway of 75 feet above the water-level having been fixed by the existing height of Runcorn Bridge, and the clear span of 120 feet by the bottom width of the canal.

The use of steel in place of wrought iron, which is a great advantage for all movable bridges in reducing the weight to be moved, is specially valuable for swing-bridges, owing to the reduction in weight that can be effected on both arms of the balanced bridge.

**Bascule Bridges.** These bridges are opened by raising the outer extremity which revolves on a horizontal hinge, or axis, placed on the abutment. The ancient drawbridges of fortresses were specimens of this class of bridge. The larger bascule bridges are composed of two portions revolving on an axis on each abutment, and meeting together in the centre of the opening. A tail-end balances the portion spanning the opening; and this tail-end sinks into a pit when the bridge is raised. If, however, the water-level is near the roadway, the pit must be made watertight to prevent the influx of water injuring the tail-end, and diminishing by immersion its effective weight as a counterpoise to the other portion.

In Holland, large overhead beams, resting upon wooden supports, counterpoise the wooden· drawbridges, and raise them by means of connecting chains. Numerous examples of these bridges may be seen at Amsterdam, where they provide roadways over the intricate network of canals which extends throughout every part of the city. Iron bascule bridges have been erected for crossing navigations and locks ; but they are rarely used in preference to swing-bridges, for they are not convenient for large spans.

The central span of the Tower Bridge over the Thames, below London Bridge, is by far the largest bascule bridge hitherto constructed ; it has two counter-balanced leaves, meeting in the centre to form an arch, spanning an opening of 200 feet; and the two leaves are raised by hydraulic machinery to a vertical position within a recess in the face of each tower, or lowered to form the bridge, 50 feet in width [1]. The bridge is mainly composed of steel; and each leaf with its counterpoise weighs 950 tons. The inconvenience of swinging a long bridge round in the middle of a river crowded with traffic, the desire to provide an uninterrupted passage for foot-passengers, and æsthetic considerations, led to the adoption of a bascule bridge in this special case, where the high towers, giving scope for architectural effect, hide the leaves of the central span when raised, and give access and support to a permanent high-level foot-bridge.

**Lift-Bridges.** The superstructure of a lift-bridge is raised vertically to a sufficient height to afford the requisite headway for vessels to pass underneath. This form of movable bridge has been rarely adopted, for it involves lifting bodily the whole weight of the superstructure to the necessary height, and does not leave the passage perfectly free like a swing-

---

[1] 'Achievements in Engineering,' L. F. Vernon-Harcourt, pp. 153–156, and illustration.

bridge or a bascule bridge; but, on the other hand, it occupies very little space on the quays, and in this respect compares very favourably with swing-bridges. A lift-bridge was erected in 1849, for carrying a branch of the Brighton Railway over the Surrey Canal, owing to the impossibility of purchasing sufficient land to erect a swing-bridge. The bridge was lifted by wire ropes fastened to its four corners, which passed over grooved wheels supported on cast-iron standards.

Two lift-bridges have been erected over the Ourcq Canal at La Villette in Paris, one in 1868, and the other in 1885. The latter, which is considerably the largest, carries the rue de Crimée over a passage between the two basins of La Villette in communication with the Ourcq Canal, and replaced a swing-bridge when the passage was enlarged in width from 25½ feet to 50 feet. This lift-bridge has a total length of 59 feet, a width of 25½ feet, and weighs about 83½ tons ; and it is counterbalanced by weights hung from four chains passing over pulleys carried by four cast-iron columns, the chains being attached to the four corners of the bridge [1]. It is lifted the full height of 15 feet in one minute, by two hydraulic presses placed at each extremity of the axis of the bridge.

In the United States, two lift-bridges cross the Oswego Canal at Syracuse, and are lifted by hydraulic power, the employment of which is expedient for working such bridges when the span is large and the traffic across the bridge considerable. The largest lift-bridge, however, hitherto erected, has been recently completed at Chicago, carrying the traffic of Halsted Street across the Chicago River [2]. This bridge, of 130 feet span, is lifted 155 feet above the water-level of the river, by steel wire ropes running round wheels supported by a superstructure resting on two light steel-braced towers, about 200 feet high, situated at each extremity of the bridge, and guiding its motion. The lift-bridge, with its

[1] Annales des Ponts et Chaussées, 1886 (2), p. 709, and plate 20.
[2] 'Engineering News,' New York, April 19, 1894.

counter-balancing weights, ropes, and chains, weighs 600 tons, and is readily lifted by a steam-engine to its full height, enabling masted vessels to pass under it. The bridge carries a roadway 36 feet wide, and footways on each side, 7 feet in width.

**Aqueducts for Canals.** Aqueducts carry canals over roads and railways, and across valleys and rivers; and they resemble ordinary bridges and viaducts, with the exception that the superstructure has to be made suitable for containing the water channel, and the side walls of the superstructure, taking the place of parapets, have to be made strong enough to resist the pressure of the water. One of the side walls carries the towing-path; and the water channel through the aqueduct is often contracted to the width for a single barge. These aqueducts have to support the constant and uniformly distributed load of the water, instead of the rolling loads which railway and roadway bridges have to bear; but in other respects their design and construction are similar to ordinary railway bridges and viaducts.

Occasionally, when a canal has to cross a river, the construction of an aqueduct is dispensed with, by making the vessels navigating the canal use the river for crossing to the opposite side; but unless the river has a steady flow, and a fair depth at all seasons, the canal traffic is exposed to serious difficulties and delays in flood-time and during dry weather. The lateral canal of the Loire, on the left bank of the Loire, communicated till recently with the Briare Canal on the right bank, by a passage across the river which had been specially trained near Briare to minimize the difficulties of transit and provide an improved depth. Nevertheless, the high floods, rapid currents, shifting bed, and deficiency of water at a low stage, rendered the passage of the river dangerous at times, and generally inconvenient, in spite of the provision of a special tug for assisting vessels in the transit since 1881. Accordingly, owing to the importance of this passage, which forms a link in the much-frequented waterway con-

necting the Loire basin with that of the Seine, the transit
of the river has been suppressed by the construction of an
aqueduct across the river at Briare, connecting the two
canals. This aqueduct, recently constructed, is the largest
canal aqueduct in France : it consists of a steel superstructure
resting upon masonry piers, and crossing the Loire in fifteen
spans of 131 feet; and its total length, including a span of
27 feet across a branch of the lateral canal, amounts to 2,174
feet ; whereas the Agen aqueduct, the next longest in France,
is 1,767 feet in length. The spans, moreover, of the Briare
aqueduct exceed those of any previous aqueduct in France by
28 feet. The width between the main girders of this aqueduct
is 24 feet, and the depth of water is $7\frac{1}{2}$ feet; whilst towing-
paths, $8\frac{1}{8}$ feet wide, are supported on projecting brackets
fastened to the main girders. The estimated cost of the Briare
aqueduct, exclusive of numerous auxiliary works for connecting it with the canals on each bank, was £113,500. Stop-
gates are provided at the ends of the aqueduct for shutting it
off from the canals on each side in the event of an accident ;
and eight sluice-gates in the abutments enable the aqueduct
to be rapidly emptied.

Streams flowing across the line of a canal must be spanned
by an aqueduct, or carried under the canal by a culvert or an
inverted siphon. The waterway provided for these streams
must be adequate to discharge freely the maximum flow of the
stream, as any damming back of the flood of a stream would
compromise the security of the canal.

The Barton aqueduct carrying the Bridgewater Canal across
the river Irwell at a height of 39 feet, constructed in 1760–
61, was the first canal aqueduct built in England ; and now the
Barton swing aqueduct, recently erected close by, carries the
same canal across the Manchester Ship-Canal, and furnishes
the first example of a movable aqueduct, though previously
similar troughs containing water have been raised vertically
in canal lifts at Anderton on the river Weaver, and elsewhere.

**Tunnels for Canals.** The tunnels through which canals are sometimes carried, especially at their summit-level in passing from one river basin to another, are precisely similar in construction to the tunnels which have been so frequently made for railways, with the exception that tunnels for canals must necessarily be made level from end to end, instead of with the gradients which are introduced into railway tunnels to provide for drainage or differences in level at the two ends. The dimensions of the tunnel have to be regulated by the depth of water, the headway required for the barges or vessels, and the widths of the channel and towing-path. The width of the waterway through a tunnel is often restricted so as to admit of the passage of barges in one direction only at a time, for the sake of economy in construction. Sometimes, also, the towing-path is dispensed with to reduce the size of the tunnel; and the passage through the tunnel is effected by the boatmen themselves pushing against the roof of the tunnel by poles or their feet, or by hauling on chains fastened along the sides of the tunnel, or by steam towage or rope traction.

The earlier tunnels were constructed for canals, the Hare-castle tunnel at the summit-level of the Trent and Mersey Canal, $1\frac{3}{4}$ miles long, having been constructed in 1766–77, only 12 feet high and $9\frac{1}{4}$ feet wide. The length of time occupied in the construction of this tunnel was due to the want of experience in such work at that period, for a second tunnel, parallel to the first and of the same length, 16 feet high and 14 feet wide, completed in 1827 to provide for the increase in traffic, was executed within three years. There are nine other tunnels on the canals of Great Britain of over a mile in length [1], the longest being the Standedge tunnel on the Huddersfield Canal, piercing the ridge separating the valleys of the Tame and the Colne, $3\frac{1}{10}$ miles in length, and parallel to the tunnel of about the same length through which the railway from Man-

---

[1] 'Returns made to the Board of Trade, in respect of the Canals and Navigations in the United Kingdom, for the year 1888.'

chester to Huddersfield now runs. The second longest canal
tunnel in England is the Sapperton tunnel on the Thames and
Severn Canal, passing through the ridge separating the basins
of these rivers, $2\frac{1}{8}$ miles long; and this ridge is also crossed,
at a different point, by the Great Western Railway between
Kemble Junction and Stroud, but in a shorter tunnel owing
to the greater elevation attained by the railway.

On the French canals, there are five tunnels exceeding one
mile in length, all situated at the summit-level of canals connecting two river basins. The longest of these tunnels is on
the St. Quentin Canal connecting the basins of the Scheldt
and the Somme, having a length of $3\frac{1}{2}$ miles, 21 feet wide, and
$13\frac{3}{4}$ feet available height; the second longest tunnel is on the
Marne-Rhine Canal, with a length of 3 miles; and the third,
on the Burgogne Canal joining the Rhone and Seine basins, is
2 miles long [1]. Tunnels are not required on lateral canals
following the valleys of rivers; and some canals even of considerable length, and traversing the ridge between two river
basins, have been constructed without tunnels. Thus the
Canal du Centre connecting the basins of the Saône and the
Loire, 72 miles long, has no tunnel on it; and the Canal du
Midi, 150 miles long, rising to a height of 600 feet above the
sea at its summit-level, and passing from the basin of the
Garonne across the basins of the Aude, the Orb, and the
Hérault, is without a tunnel.

The tunnels constructed for railways are, at the present day,
far more numerous than tunnels for canals; whilst the longest
tunnels, and the chief progress in tunnelling belong to railway
engineering. Thus there are numerous railway tunnels, in
England alone, of over a mile in length; whilst the Severn
tunnel, the tunnels through the Alps at the Mont Cenis,
St. Gothard, and Arlberg [2], and the Hoosac tunnel in the United

[1] 'Guide Officiel de la Navigation Intérieure.' Paris, 1891.

[2] 'Alpine Engineering,' L. F. Vernon-Harcourt, Minutes of Proceedings Institution C.E., 1889, vol. xcv. pp. 257–271.

States, are considerably longer, and were more difficult works than any canal tunnels. Accordingly, the experience gained in these works can be utilized with advantage in the construction of tunnels for canals.

**Waste Weirs, and Sluices.** When a canal is fed by a river or watercourses, it is necessary to provide some means of regulating the supply of water, otherwise an over-abundant supply in flood-time might cause the water to overtop the banks, both stopping the traffic and injuring the banks and the adjacent land. Accordingly, waste weirs have to be constructed at suitable places, over which the water is discharged into watercourses, or into cuts leading to adjacent streams, when it rises above a certain level. The level is fixed by the height given to the sill of the weir, and the discharge of water by its length.

Sluices, or draw-door weirs, are also serviceable for emptying the canal when required; and they become necessary when a large volume of water has occasionally to be discharged. Sluice-gates working on free rollers have been provided for controlling the passage of the waters of the Mersey and the Irwell, through the sluices alongside the locks of the Manchester Ship-Canal; and similar sluices have been constructed along the canal in front of the mouth of the Weaver, to regulate the discharge of this river, and also alongside the Eastham locks to adjust the water-level in the tidal reach of the canal. The surplus water, moreover, flowing into the canal at spring tides, is discharged through the tidal openings left at places in the river embankment of the canal along the tidal reach, acting as waste weirs, supplemented by the Weaver sluices, and another set of sluices at Old Randles between Runcorn and Warrington.

**Stop-Gates.** As canals contain large volumes of water above the level of the low-lying lands traversed in places by the canal, it is important to take precautions against the extensive flooding of the adjacent land, and the laying dry of a long reach of a canal, which the sudden failure of a canal bank might entail. Long canal reaches, especially where situated on embankments, are, consequently, divided into several sections by stop-gates, which either consist of a pair of gates similar to lock-gates, or of a series of planks or panels let down between grooves in vertical walls on each side of the canal. These walls are either constructed specially for the stop-gates, or the abutments of bridges over the canal or the parapets of aqueducts are utilized for the purpose. This division of the canal into comparatively short sections is, moreover, valuable in the event of any repairs being needed, as thereby only a portion of the water in the reach has to be drawn off.

**Supply of Water for Canals.** An adequate supply of water must be provided for a canal, to make good losses from leakage, evaporation, and expenditure of water in locking. Frequently this supply can be furnished by adjacent streams, or by the main river in the case of a lateral canal, or sometimes by a lake, springs, artesian wells, or pumping ; but when a constant adequate supply is not procurable from these sources, which is especially liable to be the case at the summit-level of a canal, an artificial reservoir must be formed, at a sufficient elevation, by placing a dam across the valley of a watercourse, which being filled in the winter, is able to furnish the necessary quantity of water to the canal throughout the dry season.

**Reservoirs.** The largest reservoir dams, for impounding water in artificial reservoirs in the valleys of rivers, have been constructed for the water-supply of towns and for irrigation, to which branches of engineering the formation of reservoirs mainly belongs [1]. Nevertheless, many canals are partially supplied, particularly in their upper reaches, by numerous reservoirs which have been mostly formed by means of earthen reservoir dams placed across the valleys of small rivers at

---

[1] 'Water-Supply,' L. F. Vernon-Harcourt, 'Encyclopædia Britannica,' ninth edition, vol. xxiv. p. 406.

suitable points. Thus the Grand Junction Canal is fed by eight reservoirs along the Tring summit-level, and by two reservoirs at the Braunston summit-level, having a total capacity of $7\frac{1}{2}$ million cubic yards. The Birmingham Canal Navigations are supplied by five reservoirs, with a total capacity of $6\frac{3}{4}$ million cubic yards; and the Leeds and Liverpool Canal is supplied by six reservoirs with a capacity of $5\frac{1}{4}$ million cubic yards, and the Rochdale Canal Navigation by seven reservoirs with a capacity of $9\frac{1}{2}$ million cubic yards. The Crinan Canal draws its supply from nine reservoirs having a capacity of 5 million cubic yards, and the Forth and Clyde Navigation from six reservoirs having a capacity of 9 million cubic yards; whilst the Caledonian Canal is fed by the lakes through which it passes, and the Ellesmere Canal by Bala Lake as well as by two reservoirs.

Similarly, in other countries, the supplies of water required for canals traversing the water-parting of two adjacent river basins, have necessitated the construction of reservoirs; whilst in France, the completion of the lines of inland navigation, and the enlargement of the main lines of existing canals have involved the formation of new reservoirs. Thus the Canal du Centre is supplied with water from twelve reservoirs, ten of which feed the summit-level, with a total capacity of $28\frac{1}{2}$ million cubic yards, nine of them having been constructed at the same time as the canal; whilst the reservoirs of Montaubry, Le Plessis, and Torcy-Neuf were constructed in 1851-60, 1868-70, and 1883-87 respectively, the last reservoir being considerably the largest, with a capacity of nearly $11\frac{1}{2}$ million cubic yards[1]. Earthen dams were constructed for the formation of all these reservoirs. The Burgogne Canal is fed by six reservoirs having a total capacity of $41\frac{1}{2}$ million cubic yards, three of which supply the summit-level, and two supply the canal in its descent to the Saône; whilst the sixth, which was constructed in 1878-81, supplies the canal within the

---

[1] 'Les Réservoirs d'Alimentation, Canal du Centre et Canal de Bourgogne,' A. J. B. Fontaine, V^mo Congrès International de Navigation intérieure. Paris, 1892.

Yonne basin.   Only two of these reservoirs have been formed by means of earthen dams, masonry dams having been adopted for the other four, which, however, with the exception of the recent dam for the Pont reservoir, have sections which do not accord with modern theory as illustrated by the Furens and Ban dams.   Probably the oldest reservoir of importance for supplying a canal is that of Saint-Ferréol, commenced in 1667 with the object of supplying water to the summit-level of the Canal du Midi at Naurouse, which was formed by constructing an earthen dam, with a central wall of masonry, across the valley of the river Laudot, supporting a maximum head of water of 103 feet, the reservoir when full having an area of 165 acres and a capacity of $8\frac{1}{4}$ million cubic yards [1].   The Marne-Saône Canal, in course of construction, will be supplied with water in dry seasons, for a length of 118 miles, by four reservoirs with a total capacity of $58\frac{1}{2}$ million cubic yards, two of which have been already completed, one by means of an earthen dam 54 feet high, and the other by the erection of a masonry dam with a maximum height of 101 feet [2].

Reservoirs have been formed in Russia by wooden dams of small height, for supplying water to the summit-levels of some navigations, such as the Marie Canal connecting the basins of the Volga and the Neva, and the Dnieper-Bug Canal.   The principal reservoirs, however, in Russia have been formed by dams across the upper portions of rivers, closed on the melting of the snow to provide water for navigation in the river below during the summer [3].   These dams, which constitute in reality draw-door weirs, are generally made entirely of wood, though sometimes masonry piers have been introduced, and the head of water retained does not exceed $17\frac{1}{2}$ feet; but owing to the

---

[1] 'Les Réservoirs dans le midi de la France,' Marius Bouvier, Vme Congrès International de Navigation intérieure.   Paris, 1892, pp. 21–24.

[2] 'Les Réservoirs du Département de la Haute-Marne,' Gustave Cadart, Vme Congrès International de Navigation intérieure.   Paris, 1892.

[3] 'Les principaux Réservoirs établis en Russie,' Émile Hoerschelmann, Vme Congrès International de Navigation intérieure.   Paris, 1892.

flatness of the country, the amount of water thus stored up
is very large, attaining a maximum of 500 million cubic yards
in the Verknévoljsky reservoir on the Volga, supplying the
Upper Volga with water from April to October for a distance
of 450 miles. The draw-doors regulate the supply of water
to the river below ; and in the autumn, the weirs are opened
to their full extent till the approach of the spring, when it is
necessary to replace the draw-doors with caution to prevent the
weir being injured by floating ice coming down against it.

**Reservoir Dams.** Earthen dams have generally been constructed in forming reservoirs for canals, but masonry dams
have also been occasionally adopted. A watertight embankment of earthwork is expedient for a dam of moderate height,
in a moist climate, especially where an impermeable stratum
suitable for the foundation of the dam, but not consisting of
rock, is found a short depth below the surface, and good materials for forming the dam can be readily obtained. A masonry
dam is preferable for dams of considerable height, where
a foundation of rock is available. All loose material must in
every case be removed from the site of the dam.

The embankment forming an earthen dam is generally
composed of a central wall of puddled clay, carried down into
an impermeable stratum, supported on each side by earthwork
of suitable materials brought up in thin layers carefully rolled.
The inner slope towards the reservoir is usually made 3 to 1,
and is protected by clay, pitching, or concrete from the wash of
the waves ; the outer slope is formed to the angle at which the
material employed will readily stand, ordinarily between 2 to 1
and 3 to 1 ; and berms are sometimes introduced at intervals
to provide against slips. Sometimes the impermeability of
the embankment is effected by lining the inner slope with
a layer of concrete in place of the central puddle wall, any
infiltration being provided for by a layer of rubble down the
back of the concrete lining, leading the water to the outlet
culvert ; and occasionally the material of the embankment is

sufficiently retentive to enable the puddle wall to be dispensed with. An embankment, however, formed wholly of clay is liable to crack in dry weather, leading to slips in wet weather and after frosts; and materials approximating in composition to two of sand and one of clay have been found most suitable for the mass of the embankment. The outlet culvert is generally laid in the solid ground, under the centre of the embankment; or sometimes it is placed in a tunnel pierced through the side of the valley, which secures the dam against the serious injury resulting from leakage through the culvert when fractured by unequal settlement under the weight of the superincumbent embankment. When the culvert passes under the embankment, it should invariably be controlled at its inner extremity by a valve tower, with valves for drawing off the water at various levels from the reservoir, as adopted at Torcy-Neuf reservoir, which enables the culvert to be shut off from the reservoir for inspection and repairs. A bye-wash is provided at a suitable place at the side of the embankment, with the level of its sill, and its width, so adjusted as to discharge readily any flood waters from the reservoir, and so secure the dam from any danger of being overtopped by the water in the reservoir.

Masonry or rubble concrete is generally adopted for dams exceeding 60 to 80 feet in height, on a solid rock foundation. A masonry dam is given a very slight batter on its inner face, the considerable curved batter on its outer face providing the rapid increase in width downwards required to sustain the increased pressure towards the base, resulting from the weight of the dam, and the pressure of the water, in proportion to the depth, with the reservoir full. The lines of resultant pressures, with the reservoir empty and full, should be within the middle third of the width of the dam, to avoid the possibility of tensional strains; and the pressures should at no point exceed the limit which the masonry can readily bear without injury.

The Furens and Ban dams in France were the first high masonry dams designed on correct theoretical principles. The Furens dam, constructed in 1861–66, retains a maximum depth of water of 164 feet; it is 18¾ feet wide at the highest water-level of the reservoir, and 161 feet at its base; and it was designed so that the pressures should nowhere exceed 6·1 tons on the square foot. The Ban dam, built in 1866–70, and retaining a maximum depth of water of 148 feet, is 18 feet in width at the highest water-level, and 122 feet at its base; but in this dam, the maximum limit of pressures was raised to 6·6 tons per square foot, and higher limits than this are considered admissible. The outlet culvert is usually formed in a tunnel through the rock in the slope of the valley at one side of the dam; but occasionally it is constructed through the dam, as for instance in the Ternay dam in France and the Villar dam in Spain. The bye-wash is generally provided at the side; but in the Vyrnwy dam, the surplus water flows over the dam, under the arches carrying the roadway over the dam.

**Introduction of Supply of Water.** The supply of water should not be all provided at the summit-level, as, with the exception of the expenditure of water in the passage of boats, the causes of loss are equally distributed throughout the canal. Moreover, except in the case of lateral canals, greater difficulty is experienced in providing an adequate supply of water at the summit-level of a canal than along the lower reaches. Accordingly, supplies of water should be furnished at suitable points along the canal, so that the losses may be made good as near as possible to the places at which they occur.

If, on the contrary, the supply is all derived from the summit-level, it is very difficult to regulate the supply which must be passed through the lock sluices so as to meet the varying demands. Wherever watercourses form the source of supply, their minimum discharge must be taken as the

basis in estimating the amount that can be derived from
them.

**Consumption of Water in Canals.** The waste of water
per day, by leakage in the channel and by evaporation.
has been estimated, for the canals in Great Britain. as
equivalent to a depth of two inches over the whole surface
of the canal ; and the loss by the flow of water through
the canal, due to leakage at the lock-gates, at between
10,0co and 20,000 cubic feet per day. The loss of water.
however, from evaporation depends upon the climate and the
season of the year ; and the loss from leakage in the channel
depends upon the state of the works, the nature of the
soil, the depth of water, and also the age of the canal, for if
the canal is in good condition, the channel becomes gradually
more watertight by the deposit of silt. In France, for
canals with a surface width of about 50 feet, the loss per
day from evaporation and leakage has been estimated at
a minimum of about 700 cubic yards per mile for old
canals in good condition, rising to 2,000 cubic yards per
mile for canals of more recent construction, and attaining
sometimes more than treble that amount in new or reconstructed canals.

The consumption of water in locking depends upon
the size of the locks, and the amount of traffic through
them. Each operation of locking withdraws a lockful of
water from the upper pool in three out of the four possible
cases, the exception being when a descending boat finds
the lock full. The least consumption of water is attained,
in the case of a single lock. when single boats ascend and
descend alternately, a single lockful being under these
circumstances sufficient for two boats. In a flight of
several locks, however, the locking of a train of boats up
and down causes much less expenditure of water than an
equal number of boats ascending and descending alternately.
Accordingly, a flight of locks is sometimes made double,

one flight being used for ascending, and the other for descending boats, whereby the consumption of water is considerably lessened. When the supply of water is very limited, the water expended in locking is sometimes pumped up again into the upper reach; and often expedients have been resorted to for reducing the loss of water resulting from a large traffic, which will be described in the next chapter.

The emptying of the canals for the execution of repairs during the annual stoppage of traffic, which commonly takes place for a month in the summer on the French canals, necessitates the supply of a considerable volume of water for filling the canals again before they can be reopened for traffic, at a period of the year when the supply of water is liable to be getting low after a dry season.

**Remarks on Canal Works.** The laying out of the route of a canal, and the earthworks in the construction of a canal involve much greater care, and a larger cost, than similar works for a railway, on account of the necessity of forming long level reaches for a canal, and providing and maintaining a watertight channel throughout. Moreover, the provision of an adequate supply of water in the upper reaches of a canal necessitates important auxiliary works, such as reservoir dams and conduits; whilst at every change of level, the lock, incline, or lift required for transferring vessels from one reach to the other, involves a considerable cost, depending on the size of the vessels to be accommodated and the difference in level between the reaches. The bridges, aqueducts, tunnels, and culverts incidental to the construction of canals, are very similar to the works required in the formation of railways; and the engineers of the canals constructed in the eighteenth, and the earlier portion of the nineteenth century, acted as pioneers in these works, and greatly facilitated the subsequent development of railways which eventually took the lead in the extension of such works. The adoption of tunnels on railways mainly depends upon the saving of cost in earthwork; but on canals, the saving of water and the reduction in the number of locks, in traversing the summit-level, have also to be considered. In these respects, tunnels would appear to be more expedient for canals than for railways; but the great superiority of railways for traversing rugged country, with their adjustable gradients and sharp curves, have enabled them to penetrate regions wholly unsuited for canals; and the longest railway tunnels, such as the Alpine tunnels and the Severn tunnel, have been constructed in places quite inaccessible for canals.

Inland navigation canals are, indeed, best suited for low-lying, flat countries such as Holland, parts of Belgium, the adjacent northern corner of France, and the neighbourhood of Venice, where abundance of water and very slight differences of level enable them to attain a high state of development at a comparatively small cost. In fact in many parts of Holland, and in Amsterdam and Venice, canals form the main arteries for goods traffic, and have in a great measure superseded roads and streets. Canals also which connect long lines of river or lake navigations, are very valuable by greatly enhancing the importance of the navigations which they unite, and thereby rendering services quite disproportionate to their actual length. Moreover, where a large traffic in bulky goods has to be provided, canals may be advantageous in districts not wholly favourable for their construction; but in hilly or rugged country, the essential condition of level reaches precludes the possibility of canals competing successfully with railways.

# CHAPTER XVI.

## CANAL LOCKS.

Locks for surmounting Differences in Level on Canals. *Sizes of Canal Locks* :
Accommodation for Vessels, enlargement of Chamber, intermediate Gates, double
Locks; Standard Dimensions, in Great Britain, in France, and in America.
Enlargement of Sault-Sainte-Marie Canal Locks. Dimensions of Ship-Canal
Locks. *Lifts and Flights of Locks* : Instances in United Kingdom, and in France;
Advantages of Flights; Reduction in Number of Locks, and Increase of Lifts, on
Canal du Centre, on St. Denis Canal. Lock with Lift of 32·5 feet, Description.
Lifts of Ship-Canal Locks. *Reduction of Time in Locking* : Deepening of Lock-
Chamber, and Longitudinal Sluice-Ways; Cylindrical Sluice-Gates, on River
Weaver, and in France; Summary of Arrangements for Saving Time. *Saving
Water in Locking* : Side Ponds; Double Locks; Contrivances at Aubois Lock.
Remarks on Canal Locks.

LOCKS are the means commonly employed on canals for
raising or lowering vessels from one reach to another. These
canal locks are worked in the same manner as locks on rivers,
described in Chapter V ; and they are similar in construction
to river locks, with the exception that a lift-wall, serving as
a retaining wall where the excavation is stepped down, is
built at the upper end, upon which the upper sill rests, so that
the upper pair of gates are less in height than the lower pair
of gates by the amount of the lift of the lock (Fig. 11, p. 101,
and Plate 11, Fig. 9). The greater variations in level, however,
of the ground generally traversed by canals than experienced
along the lower parts of river valleys, especially where canals
have to connect rivers by crossing the water-parting of their
basins, involve the more frequent introduction of locks, and
consequently more repeated delays in passing through the
locks, and a greater expenditure of water in locking. Accordingly, special care has to be directed, in the design of canal

locks, to the facilitating of the rapid passage of vessels through them, and to the reduction of the expenditure of water in locking, more particularly as the water-supply for canals has to be artificially obtained.

There are four points which specially need consideration in designing locks for canals, namely, (1) the size of the locks ; (2) the lift of the locks ; (3) the reduction, as far as possible, of the time occupied in locking ; and (4) the saving of water expended in lockage.

## Sizes of Canal Locks.

The dimensions of the locks on any canal should be adequate to admit the largest vessels navigating the waterways with which the canal is connected.

**Arrangements for Accommodating Vessels.** Where the traffic is generally conducted by trains of barges, the locks are made large enough to receive several barges, by widening out the lock-chamber between the entrances, or by lengthening the lock and introducing an intermediate pair of gates. The latter system provides for the passage of single vessels with less expenditure of water ; but the first plan is very commonly adopted, and is cheaper in construction in proportion to the water area of the lock-chamber obtained. When vessels of very different sizes have to be accommodated, locks of different dimensions are occasionally built side by side, thereby both saving time and water in locking the smaller vessels singly, and increasing the facilities for the passage of vessels. This arrangement has been adopted at both extremities of the Amsterdam Ship-Canal, and also for the locks on the Manchester Ship-Canal, with an intermediate pair of gates as well in most instances (Plate 12, Figs. 3, 4, 9, and 10) ; whereas the two regulating locks at each end of the Baltic Canal, have been made exactly the same size, and provided in each case with intermediate gates in the centre of the lock-chamber (Plate 12, Figs. 14 and 15).

**Standard Dimensions of Canal Locks.** Several canal locks in England are only about 74 feet long and 7 feet wide, with 4 to 5 feet of water over their sills: whilst many of them are 14 to 16 feet in width, with a similar length and depth. The ordinary dimensions of canal locks in Ireland are 70 feet length, 15 feet width, and 5 to 6 feet depth of water on sill[1]. The Aire and Calder Navigation, however, which accommodates vessels 120 feet long, 18 feet beam, and 9 feet draught, and conveys long trains of coal barges, has thirteen locks, 201 to 339 feet long, 18 to $22\frac{1}{2}$ feet wide, and $8\frac{1}{2}$ to 10 feet depth of water on sill; whilst the Exeter Canal has one lock 300 feet long, $26\frac{2}{3}$ feet wide, and 13 feet depth of water on sill[2]. In France, the minima standard dimensions for the locks on the main lines of inland waterways, were fixed by law, in 1879, at $126\frac{1}{3}$ feet length, 17 feet width, and $6\frac{1}{2}$ feet depth of water, so as to accommodate barges of 300 tons, and a minimum headway under bridges of $12\frac{1}{2}$ feet. The locks on the Welland and St. Lawrence canals have been successively enlarged to a standard size of 275 feet in length, and 45 feet in width, so as to provide for the passage of vessels of 1,500 tons, drawing 14 feet of water, between Lake Erie and Montreal[3].

**Enlargement of Sault-Sainte-Marie Canal Locks.** Owing to some impassable rapids on the St. Mary river, which connects Lake Superior with Lake Huron, a lateral canal was opened in 1855, at Sault-Sainte-Marie on the United States side of the river, $1\frac{1}{4}$ miles long, with two locks in it to surmount the difference in level of 18 feet due to the falls, each lock being 348 feet long and 69 feet wide. By 1870, the development of the navigation necessitated the enlargement and deepening of the Sault-Sainte-Marie Canal to a width of 108 feet and a depth of 16 feet; and a single lock was built

---

[1] The lengths given for locks refer to the available length of the lock-chamber.

[2] 'Returns made to the Board of Trade in respect of the Canals and Navigations of the United Kingdom, for the year 1888.'

[3] 'The Waterways of Canada,' Sandford Fleming, Manchester Inland Navigation Congress, 1890.

alongside the two existing locks, 515 feet long, 60 feet wide
at the entrances, and 80 feet in the lock-chamber, and affording a depth of 16 feet of water on the sill, the change of level
being effected in a single lift of 18 feet.  These works, completed in 1881, have already become insufficient for the increasing demands of the navigation; and extension works are
in progress.  The canal is being given a depth of 20 feet; and
a new lock is being constructed on the site of the two original
locks, having a length of 800 feet, a width of 100 feet, and
a depth of 21 feet of water on the sill.  A lateral canal is also
being constructed with the same object on the Canadian side
of the river, with a width of 153 feet and a minimum depth of
18 feet, together with a lock 600 feet long, 80 feet wide in the
chamber, and 60 feet at the entrances, affording 16 feet of
water over the sill, and having a lift of 18 feet.

**Dimensions of Ship-Canal Locks.**  The Caledonian Canal,
constructed as a ship-canal early in the nineteenth century,
has locks 170 feet long and 40 feet wide, with a depth of 17
feet of water on the sill.  The largest lock at the North Sea
end of the Amsterdam Ship-Canal, is 390 feet long and 60 feet
wide, with an available depth of about 24 feet of water on the
sill; but a new lock in a side cut, which is approaching completion, has been made 776 feet long, 82 feet wide, and $31\frac{1}{4}$ feet
depth of water on the sill, so as to allow the largest class of
ocean-going steamers to enter the canal (Plate 12, Figs. 1
and 3).  The largest lock on the Manchester Ship-Canal, at
Eastham, is 600 feet long and 80 feet wide; and its upper sill
is 28 feet below the lowest water-level in the canal, or two feet
lower than the bottom of the canal.  The double locks at
each end of the Baltic and North Sea Ship-Canal, have been
made 492 feet long and 82 feet wide; and their sills are $31\frac{2}{8}$
feet below the mean water-level.

The small dimensions of the Caledonian Canal locks, and
the successive enlargements of the St. Lawrence, Welland, and
Sault-Sainte-Marie canal locks, show how greatly the require-

ments of navigation have increased, and indicate that it is advisable in designing such works to make the locks larger than the existing requirements, to avoid the necessity of reconstruction a few years later. It is, moreover, always expedient to place the sills of the locks somewhat lower than the general level of the bottom of the canal, so that a moderate increase in depth may be effected by dredging, without entailing the rebuilding of the locks.

## LIFTS AND FLIGHTS OF LOCKS.

The difference in level between two adjacent reaches of a canal, necessitated by the slope of the land through which the canal passes, is surmounted by the lift of the lock. When the variation in the level of the land is considerable at any point, the change of level is effected by means of a flight of locks, dividing the total lift into a series of steps.

**Instances of Lifts and Flights of Locks.** The lift of canal locks in the United Kingdom varies between $1\frac{1}{2}$ and $14\frac{1}{2}$ feet, being comprised for the most part between 4 and 9 feet. A rise of 203 feet at Tardebigge on the Worcester and Birmingham Canal, is accomplished by a flight of twenty-nine locks; a rise of 154 feet at Combe Hay on the Somersetshire Canal, is effected by a flight of twenty-two locks; whilst a flight of twenty-one locks at Wigan on the Leeds and Liverpool Canal, and a similar flight at Hatton on the Warwick and Birmingham Canal, surmount rises of $199\frac{1}{2}$ feet, and 146 feet respectively. There are also several smaller flights of locks on other canals, where abrupt changes of level are necessary in traversing hilly districts; whilst on the Caledonian Canal, there is a well-known flight of eight locks at Banavie, affording a total lift of $63\frac{2}{3}$ feet.

In France, the lift of canal locks ranges generally between $6\frac{1}{2}$ and 9 feet, and rarely exceeds 10 feet. Flights of locks have not been so frequently adopted in France as in England; and the maximum number of locks in a flight is less. There is, however, a flight of seven locks near Beziers on the Canal du Midi, and a flight of five locks at Fontinettes on the Neuffossé Canal.

**Advantages of Flights of Locks.** Locks arranged in a flight diminish the number of the gates, as compared with a similar number of single locks placed at intervals apart, since the lower gate of one lock serves as the upper gate of the next lock below it; and they reduce the cost of working and the inconvenience, by reducing the number of gates to be opened and closed, and by concentrating the work and the delay at a single spot. Unless, however, a double flight is provided, or the barges are passed through in trains, a flight of locks involves a considerably larger expenditure of water in lockage than the same number of single locks.

**Reduction in Number of Locks and Increase of Lift.** The passage of each lock entails some delay, so that the circulation of traffic is facilitated by reducing the number of the locks and increasing their lift. Certain groups of locks on the Canal du Centre of France were formerly separated by winding reaches, only 340 to 980 feet in length. When, in 1881, it became necessary to lengthen these locks from 98½ feet to the standard length of 126⅓ feet, this increase of length would have unduly reduced the length of the intervening reaches which were already too short and inconvenient to navigate, and would have aggravated the abrupt variations in the water-level in these short reaches resulting from the lockings. Accordingly, in reconstructing these locks, the lift was doubled, being increased from 8½ feet to 17 feet, which enabled one lock out of every two to be suppressed, and increased the length of the reaches to between 720 and 1,840 feet [1]. In this manner, thirteen locks were suppressed in portions of the canal with short reaches, thereby materially reducing the time occupied in transit and the working expenses,

[1] Annales des Ponts et Chaussées, 1892 (2), p. 783.

enabling the reaches to be straightened in tortuous places, and diminishing the variations in the water-level resulting from lockage.

The St. Denis Canal connects the basins of La Villette in Paris with the Seine at St. Denis ; and the improvement works on the Lower Seine, having increased its navigable depth up to Paris to 10½ feet, necessitated the deepening of the canal from 6½ feet to a similar extent, and the consequent reconstruction of the locks on it. As on the Canal du Centre, advantage has been taken of this reconstruction to reduce the number of locks from twelve to seven, by increasing their lift[1]. The locks on the canal formerly consisted of four single locks and four flights of two locks each, the lift of each lock being from 7½ to 8⅜ feet. The seven new locks have in every case been constructed with two chambers of different sizes, to accommodate the small vessels coming from the canals in the north, and the larger vessels navigating the Seine. The lifts of the four single locks have not been materially modified in the reconstruction ; but two single locks, with lifts of 14¾, and 14 feet respectively, have replaced the two flights of two locks in the lower part of the canal; whilst a single lock has been built at the Villette end of the canal, with a lift of 32½ feet, in substitution for the two adjacent flights of two locks which previously surmounted this difference of level.

**Double Lock with Lift of 32·5 Feet, St. Denis Canal** There was only one site available alongside the canal, near La Villette basins, where a lock could be constructed without stopping the traffic ; and the closing of the canal during the construction of the works was inadmissible, owing to the great importance of its navigation. The space on this site, moreover, was not sufficient for the erection of a canal lift, involving two large counterbalancing troughs side by side. Accordingly, a lock with a large lift was the only possible expedient (Plate

[1] Annales des Ponts et Chaussées, 1893 (2), p. 45.

11, Figs. 1 to 6)[1]. The small chamber has been given the standard dimensions, of 126⅓ feet by 17 feet, in all the locks ; whilst the large chamber has a length of 205 feet (constructed provisionally in some cases to only 147⅔ feet), and a width of 27 feet, to admit vessels from the Seine. The sluice-ways for both lock-chambers, with their gates and gearing, and the machinery for working all the gates, have in all cases been placed in the central pier separating the two chambers, so as to leave the roadway along the quay, on the land side of each of the two lock-chambers, perfectly free for the traffic. This has necessitated constructing the lock-gates with a single straight leaf closing right across the entrance, instead of providing two gates meeting at an angle in the centre of the span, and worked from both sides, as commonly adopted.

The difficulties to be overcome in the case of the lock with the great lift of 32½ feet, consisted in the height given by the lift to the lower gate, which, under the ordinary arrangement, would have been nearly 50 feet high, the great expenditure of water with so great a lift, and the time expended in emptying and filling a lock-chamber of such a height. The great height of gate has been avoided by building an arched bridge across the lower end of the lock, affording the standard headway adopted for the Lower Seine, of 17¼ feet above the water-level of the canal in the lower pool, and making the gate shut against the bridge at the top, thereby enabling its height to be reduced to 32⅖ feet above the sill, and diminishing the strains upon it by supporting it at the top, as well as at the sides and bottom (Plate 11, Figs. 2, 5, and 6). The expenditure of water in locking has been reduced by constructing a side pond for each chamber in the central pier, which receives a third of the water drawn off in emptying the chamber, and returns this third to the chamber in the process

---

[1] Drawings of this lock were sent to me by Mr. Guillain, Director of Public Works at the Ministry of Public Works in Paris, from which the illustrations on Plate 11 were reduced.

of filling (Plate 11, Figs. 1, 3, and 4). By means of large longitudinal sluice-ways communicating with the upper and lower pools and with the two side ponds, and connected with the chambers by a series of side outlets, the emptying and filling of the two chambers are effected as quickly at the lock with the great lift as at the others, being controlled by six large cylindrical sluice-gates for the large chamber, and four for the other (Plate 11, Figs. 1, 3, and 4).

**Lifts of Ship-Canal Locks.** The greatest lift on the Caledonian Canal is only 9⅔ feet, and on the Manchester Ship-Canal 16½ feet. Locks, however, with larger lifts than the St. Denis Canal lock, have been proposed for the Panama and Nicaragua canals. Thus in the modified design of the Panama Canal, in which it was proposed to reduce the amount of excavation by introducing locks, and thus raising the level of the canal in traversing the high ground of the isthmus, the locks were given lifts of 26¼ to 36 feet (Plate 13, Fig. 6); and in the approved scheme for the Nicaragua Canal, the proposed lifts of the locks, near each extremity, are from 29 to 45 feet (Plate 13, Fig. 10), whilst in another scheme, lifts of 50 to 100 feet were contemplated.

## REDUCTION OF TIME IN LOCKING.

By diminishing the time occupied by vessels in passing through the locks, the period of the journey from one place to another may be materially diminished, in proportion to the number of the locks, and the amount of traffic that can be accommodated along a crowded canal correspondingly increased.

**Deepening of Lock-Chambers, and Longitudinal Sluice-Ways.** The passage of each lock involves the introduction of the vessel, the closing of the gates behind it, the filling or emptying of the lock-chamber, the opening of the gates in front, and the exit of the vessel; and when the water-level in the lock is not right for the vessel, there is in addition the delay of closing the further gates and emptying or filling the lock-chamber, before the vessel can be admitted. Steam-boats can naturally pass into and out of a lock much more quickly than ordinary vessels; but the entrance and exit of all large vessels is facilitated by making the section of the lock-chamber sufficiently larger than the section of the largest boat navigating the canal, to prevent the vessel being retarded by acting like a piston on the water in the lock in entering or leaving a lock. The increased section is best obtained by making the lock two or three feet deeper than the canal, for boatbuilders are liable to make their vessels nearly the full width of the lock; whilst an increase in width of the lock augments also the consumption of water in locking. The same advantage is gained by forming sluice-ways along the whole length of the side walls, communicating with the upper and lower pools under the control of valves, and connected by numerous side openings with the lock-chamber. The raising or lowering of the water-level in the lock-chamber, which the vessel tends to produce in entering or leaving the lock, is readjusted by the efflux or influx of the water through the sluice-ways. This arrangement of sluice-ways, moreover, greatly increases the rate of filling and emptying the lock-chamber; and the provision of numerous outlets along each side prevents the vessel being driven against the gates or side walls by the rush of water. When the sluice-gates are placed in the lock-gates, it is expedient to put the upper gates deep enough for the sluice openings to be below the level of the lower pool, so as to enable the sluice-gates to be fully raised at once without creating dangerous currents in the chamber. With this arrangement, the time occupied in filling a lock is about one-third less than when the sluice openings are above the lower water-level.

Cylindrical Sluice-Gates. The raising of the ordinary, flat, vertical sluice-gates is impeded by the pressure of the water against them, greatly increasing the friction at the faces on which they slide. This impediment has been overcome by adopting cylindrical sluice-gates closing vertical circular wells communicating with the sluice-ways (Plate 11, Figs. 7, 8, and 9)[1]. The pressure of the water on these cylinders is uniform all round, and therefore the friction at the sides is eliminated, which greatly facilitates the working of these sluice-gates. Moreover, a small lift of a cylindrical sluice-gate, by exposing a cylindrical aperture, affords a considerably larger opening for the flow of the water to or from the sluice-ways, than a similar raising of a flat, vertical sluice-gate. Accordingly, these cylindrical sluice-gates are easier to lift, and they require a less amount of raising to produce their full effect; and consequently they are very valuable, in combination with longitudinal sluice-ways in the side walls, for rapidly filling and emptying large locks with a considerable lift.

Between 1874 and 1878, two locks, 229 feet long, 42½ feet wide, and having 15 feet depth of water on the sill and a lift of 8½ feet, were built on the Weaver Navigation, at Saltersford and Acton, which were each provided with six wrought-iron cylindrical sluice-gates, 5½ feet in diameter, controlling the flow of water through longitudinal culverts in the side walls and outlets at the sides, for filling or emptying the lock[2]. These sluice-gates being counterbalanced, and having the pressure equilibrated, are easily raised or lowered their full extent, of 1⅝ feet, in 4 seconds by one man.

Some years ago, in making the locks of the Scheldt-Meuse Canal the standard size, with a lift of 13 feet to reduce the number of locks (Plate 11, Fig. 7), the above arrangements were adopted, which, with the aid of hydraulic power for working the lock-gates and the sluice-gates, enabled the period occupied in locking to be reduced from between 16 and 20 minutes down to about half the period. The cylindrical sluice-gates on the Scheldt-Meuse Canal are 5¾ feet in

[1] Annales des Ponts et Chaussées, 1883 (2), p. 5.
[2] Minutes of Proceedings Institution C.E., vol. lxiii, p. 263.

diameter, and have only to be raised $1\frac{1}{2}$ feet to open the sluice-ways completely; and the filling or emptying of the lock-chamber is effected in two minutes. The same system was adopted for expediting the passage through the locks of 17 feet lift on the Canal du Centre, where two cast-iron cylindrical sluice-gates at each end of the lock, $4\frac{2}{3}$ feet in diameter, raised or lowered $1\frac{1}{4}$ feet in 12 to 13 seconds, regulate the filling or emptying of the lock, which is accomplished in $3\frac{1}{4}$ minutes; and the whole operation of locking a vessel of 150 tons, drawn in and out of the lock by two men, occupies 14 minutes, out of which only $4\frac{1}{2}$ minutes are spent in working the lock. Cylindrical sluice-gates, $5\frac{1}{4}$ feet in diameter, with large longitudinal sluice-ways and twenty-four side outlets, enable the large lock, with a lift of $32\frac{1}{2}$ feet, on the St. Denis Canal, to be filled or emptied in $7\frac{1}{2}$ to $8\frac{1}{3}$ minutes; and the large lower lock-gate, being worked by a turbine, can be opened or closed in one minute (Plate 11, Figs. 1 to 5).

**Summary of Arrangements for Saving Time.** Longitudinal sluice-ways accordingly, in the side walls of a lock, together with a slight increase in the depth at a lock, facilitate the entrance and exit of vessels approximating in mid-ship section to the width of the lock. Cylindrical sluice-gates, moreover, in combination with large longitudinal sluice-ways and numerous side outlets, ensure the rapid filling or emptying of large locks with considerable lifts; whilst the opening and closing of the lock-gates are expedited by the introduction of hydraulic power, which can also be usefully employed for working capstans to draw vessels through a lock, and thus hasten their transit.

## SAVING WATER IN LOCKING.

Where water is difficult to obtain or limited in amount, as near the summit-level of a canal and in dry weather, the saving of water in locking is of considerable importance, especially when the traffic is great, the lock-chambers large in area, and the lift considerable.

**Side Ponds.** The simplest system of saving water consists in providing one or more side ponds into which the upper portion of the water in a lock can be let off, so that the same water can be used again for refilling the lower part of the lock, instead of being discharged at once into the lower pool. This plan, for instance, has been adopted for saving some of the water discharged in emptying the lock with the great lift on the St. Denis Canal (Plate 11, Figs. 1, 3, and 4); and it is frequently resorted to where water is scarce. Theoretically, it would be possible, by constructing a series of side ponds at different levels, to save a large portion of the water; but in practice it has been found that two side ponds, economizing about half a lockful, are the most advantageous, as the loss by leakage and evaporation from a larger number of side ponds is liable to be greater than the gain.

**Double Locks.** Two locks placed side by side, with a communication between them which can be opened or closed at pleasure, effect a saving both of water and time as compared with a single lock. One lock thereby serves as a side pond to the other; and a boat enters one or other of the locks according as the water-level is the most suitable, thus preventing or diminishing the delay before entering the lock. Two flights of locks, also, placed side by side, one for ascending and the other for descending boats, economize the water used for locking.

Moreover, where the vessels vary considerably in size, an intermediate pair of gates, or two or three locks of different sizes, as previously pointed out, save water as well as time in lockage.

**Contrivances for Saving Water at Aubois Lock.** The principle of converting the *vis viva* of a current of water into a lifting force, whereby a portion of the water is raised to a higher level by suddenly altering the course of its flow, has been very ingeniously applied by Mr. de Caligny at the Aubois lock on the Loire lateral canal, so as to save a large proportion of the water discharged in locking. The flow of water through the sluice-way communicating with the upper and lower pool, a side pond, and the lock-chamber, is controlled by two wrought-iron pipes which, being raised from, or lowered on to two circular openings in the top of the sluiceway, open or close the connection between the sluice-way and the upper pool and side pond respectively [1]. By opening and closing successively the communication between the lock and the side pond in emptying the lock, some of the water rises through the pipes into the upper pool; and the subsequent free flow of the water from the lock into the side pond, raises the water in the side pond to a higher level than it has fallen to in the lock. By similarly producing and arresting the flow through the sluice-way from the upper pool in filling the lock, the water is drawn from the side pond till its waterlevel is below that of the lock; and the final filling of the lock from the upper pool raises the water-level in it sufficiently at the close to cause the upper gates to open. The lock is emptied or filled in 5 to 6 minutes; and the volume of water saved in the two operations amounts to over 60 per cent. The system, however, has not been extended to any other locks; the oscillations, which form an essential part of the scheme, are somewhat inconvenient; and though the operations can be conducted by one man, the apparatus is complicated, and notably increases the cost of the lock. The method appears more suited for special conditions, where a great saving of water in locking is essential, than for general adoption.

**Remarks on Canal Locks.** Where the differences in level between the adjacent reaches of a canal are small, and the slope of the country traversed by a canal is sufficiently

---

[1] Comptes rendus de l'Académie des Sciences, vol. lxviii, p. 118, vol. lxxxiv, p. 1213, and vol. lxxxviii, p. 362.

moderate for the changes in level to be situated some distance apart, locks furnish the simplest and best means of connecting the successive reaches. The delays, however, incidental to the passage of the locks, and the large volume of water expended in locking on a much-frequented canal, render deep chambers, longitudinal sluice-ways, cylindrical sluice-gates. machinery for working the gates and capstans, intermediate gates or locks of different sizes, and side ponds, advisable for expediting the transit and economizing the water, where the traffic is large and water is scarce.

When the declivity of the ground to be traversed by a canal is considerable, which is more or less the case in the upper portion of every river basin, locks have to be introduced more frequently, or their lift has to be increased, or they have to be grouped in flights. This frequency of locks largely increases the cost of constructing canals in rugged districts, augments the delays incurred by vessels, and necessitates a larger supply of water. Accordingly, under such conditions, special arrangements for facilitating the passage of vessels through the locks, and reducing the expenditure of water, assume enhanced importance. Where the changes in level are abrupt, and long flights of locks are introduced, the delay to the traffic, and the wasteful expenditure of water-power in the simple process of lockage become specially apparent; and, consequently, other methods have been occasionally resorted to on canals for surmounting large variations in level, which will be described in the next chapter.

# CHAPTER XVII.

## CANAL INCLINES AND LIFTS.

Inclines, and Lifts substituted for Canal Locks. *Canal Inclines:* earliest in China; Rollers for Boats; reduce Cost of Works and Time of Transit, and save Water; General Form; Methods of conveying Barges on Inclines. Barges on Wheels; Bude Canal Inclines; Shropshire and Shewsbury Canal Inclines; Morris Canal Inclines; Oberland Canal Inclines; Ourcq Canal Incline. Caissons containing Water, object; Chard Canal Inclines; Monkland Canal Incline; Georgetown Incline on Chesapeake and Ohio Canal. Ship-Railways proposed for connecting Atlantic and Pacific Oceans; Difficulties; Chignecto Ship-Railway constructed. *Canal Lifts:* compared with Inclines; Primitive Lifts on Worcester and Birmingham Canal, and on Grand Western Canal; counterbalancing caissons, lifting gates, worked by weight of water, saving of Time and Water, disuse. Anderton Hydraulic Canal Lift; object, description, working, saving of Time and Water compared to flights of locks, accident to press. Fontinettes Lift, object, resemblance to Anderton Lift, greater size, description, working, expenditure of water, cost, time of passage, settlement of press. La Louvière Lift; object, one of four proposed, similarity to Fontinettes Lift, larger dimensions, construction, weight moved, working and time of lift, cost, displacement of wall. Concluding Remarks on Inclines and Lifts.

INCLINES, and also vertical lifts, have sometimes been adopted, in preference to locks, for transferring vessels from one reach of a canal to the next, where the difference in level between the two reaches is considerable, and where a saving of time and of water in the passage of vessels is of importance. Inclines substitute a speedy transit by land on a railway, between the reaches of the canal, for the slow passage through a flight of locks; whilst lifts rapidly raise or lower vertically a movable section of the canal, in the form of an iron caisson filled with water, from one reach to another

## CANAL INCLINES.

Inclines for canals consist of a steep roadway connecting two adjacent reaches, up which barges, furnished with wheels or placed on special carriages, are drawn along rails, by ropes hauled by machinery on the top of the incline. The earliest canal inclines, of a very primitive type, were resorted to on the ancient canals of China, where, in the absence of any knowledge of the principle of the canal lock, differences in level between successive reaches were surmounted by dragging the boats up, or lowering them down a paved track, by a rope wound round a capstan. The sets of rollers placed alongside locks where a considerable number of small pleasure boats pass, on which the boats are drawn up and down, furnish examples of the smallest inclines used for waterways in the present day; but, both in principle and objects, they resemble the much larger inclines adopted on some canals.

Canal inclines, in providing conveyance by land on a gradient for carriage by water in successive steps by a flight of locks, reduce the cost of the works, save time in the transit, and dispense with the use of water, except when water-power is employed to turn the machinery for hauling up the barges. These inclines, in fact, resemble the steep inclines commonly used in mines, and occasionally introduced on railways to surmount a sudden rise in mountainous country; and one incline differs from another merely in length and steepness of slope, and in the motive power actuating the drum at the top of the incline, for winding up or unwinding the rope or wire cable attached to the barge. Two lines of way are laid on the incline, so that the ascent of one barge may be facilitated by the simultaneous descent of another. Three methods have been employed for conveying the barges on the inclines, namely, little wheels fastened under the barge, a cradle supporting the barge, and, lastly, a horizontal tank containing water, carried by a special framework on wheels, into which the barge is floated (Plate 11, Figs. 10 and 11). The first method has been adopted at the Bude Canal inclines; the second is in use on the Morris Canal inclines and elsewhere; and the third has been resorted to for the Blackhill and Georgetown inclines.

**Barges on Wheels for Inclines.** This system has a very
limited application, for it can only be used for small flat-
bottomed boats, such as those which navigate the Bude
Canal, which are 20 feet long, 5½ feet wide, and 20 inches
draught. The wheels, moreover, increase somewhat the
draught of the barge.

**Bude Canal Inclines.** Except in the lower part of the
canal near Bude, the changes of level on the Bude canal
are all surmounted by inclines, up which the small barges are
hauled by chains worked by water-power. The barges are
drawn up in trains, and are to some extent counterpoised
by the descending trains of barges. The water-power is
supplied either by means of a water-wheel turned by a stream
of water, or by two large tubs descending and ascending
alternately in two wells, the tub at the top of its well being
filled with water, and in its descent drawing the barges up the
incline. When the tub full of water reaches the bottom of
the well, the water is emptied through a flap door in the
bottom of the tub; and the empty tub in the other well,
having been drawn up its well by the descending tub, is at
the top ready to be filled with water in its turn.

**Cradles on Wheels conveying Barges on Inclines.** The
necessity for employing special barges fitted with wheels on
canal inclines, can be dispensed with by providing cars, or
cradles, for conveying the barges over the inclines. Barges of
the ordinary types can be floated on to the cradles, immersed
in the water, at one end of the incline ; then drawn out of the
canal and hauled over the incline, resting on the cradles; and
finally launched again on the canal at the other end.

**Inclines on the Shropshire and Shrewsbury Canals.** There
is an incline on the Shropshire Canal, 900 feet long and
rising 213 feet, over which barges are conveyed on cars
running on up and down lines of railway. The barges,
however, carried on this incline are only 20 feet long, 6 feet
wide, and 2¾ feet draught, little larger than the barges on the

Bude Canal ; but the available draught is increased by dispensing with wheels under the barges. Barges very similar in size, but with six inches less draught, are carried over the Trench incline of the Shrewsbury Canal, 681 feet long and rising 73½ feet.

The small size of the barges conveyed on these inclines, is accounted for by the short lengths of canals with which they communicate, constructed towards the close of the eighteenth century; whilst the Bude Canal inclines, constructed early in the nineteenth century, enabled a small canal to be carried through a hilly country, to serve a purely agricultural district.

**Morris Canal Inclines.** When the Morris Canal was constructed in 1825–31, to connect the Hudson River at New York with the Delaware River at Philipsburg, it had to be carried across a spur of the Alleghanies, with a summit-level of 914 feet above mean-tide level in the Hudson, and 760 feet above the water-level of the Delaware ; and in order to surmount these differences of level, twenty-three inclines were introduced, in addition to twenty-three locks. These inclines have gradients of 1 in 10 to 1 in 12 ; and their rise varies between 44 and 100 feet, the total rise of the inclines on the two slopes amounting to 1,448 feet. The barges, 79 feet long, 10½ feet wide, and of about 70 tons capacity, are carried on eight-wheeled cradles running on steel rails laid to a gauge of 12⅓ feet, and provided with brakes ; they are hauled up the incline by two wire ropes winding on a drum turned by a water-wheel ; and they are lowered down the incline under control of the brakes, being checked also somewhat by the water-power when laden. The launching of the barge is effected by making the first set of double-flanged wheels, on reaching the water, run upon rails, laid to a slightly different gauge, with a gentle gradient ; whilst the hind set of wheels continue to descend the steep gradient, so that the barge attains a horizontal position when fully immersed.

A loaded barge, with its carriage, weighing altogether about
110 tons, is drawn up an incline having a rise of 51 feet, in
$3\frac{1}{2}$ minutes ; and as a barge travels up or down the inclines
at the average speed of towage, the whole of the delay caused
by a flight of locks is saved. The expenditure of water in
conveying a loaded barge on an incline, has been estimated
to be only one twenty-third of that required for a flight of
locks of the samel ift.

**Oberland Canal Inclines.** Four inclines were constructed
on the Oberland Canal in Prussia in 1844–60, with a gradient
of 1 in 10, and rises of 66 to $80\frac{1}{3}$ feet, up which barges
carrying 70 tons are drawn on iron cradles running on steel
rails laid to a gauge of $10\frac{3}{4}$ feet. The cradles are supported
on two four-wheeled bogies, 30 feet apart, which are capable
of turning on a horizontal axis to adjust themselves to
differences of slope. The total load, including the weight of
the barge and cradle, is 105 tons, which is hauled up by a
wire rope, worked by a water-wheel supplied with water from
the canal, at a speed of about 3 feet per second, so that the
longest incline is traversed in about $4\frac{1}{2}$ minutes. A similar
arrangement is provided for launching the barges horizontally
as on the Morris Canal. Another similar incline has been
constructed more recently to take the place of the five lowest
locks on the canal.

**Ourcq Canal Incline.** The Ourcq Canal approaches within
two-thirds of a mile of the river Marne, at Beauval near
Meaux ; but formerly it was necessary to make a round of
about 60 miles to get from the canal up to this point on the
Marne by water. The water-level of the canal at this place is
40 feet higher than that of the Marne, so that a flight of four
or five locks would have been required to connect them by
water ; and these locks would have expended more water than
the canal could supply. Accordingly, the connection between
these two waterways has been effected by an incline of 1 in
25, up which barges, 92 feet long, 10 feet wide, and 4 feet

deep, weighing when loaded 70 to 75 tons, are drawn on a wrought-iron cradle supported on a pair of four-wheeled bogies running on rails laid to gauge of 6⅓ feet [1]. The cradle, which is 79 feet long, and weighs 35 tons, is drawn up by means of a wire rope worked by a turbine on the Marne, which turns a cog-wheel carried by the cradle, and moving in a rack placed between the rails. The cog-wheel and rack were added in 1888, to ensure steadiness of motion. The wheels of the cradle have a central flange enabling them to run on an outer-or inner track, so that by slightly altering the gauge and inclination of an additional inner or outer line of rails, the front bogie on dipping into the canal, or the hind bogie on descending into the river, may change its gradient, and bring the barge into a horizontal position for launching.

This incline, with its summit-level carried slightly above the water-level of the canal, and its dip on the other side into the canal, has a total length of 1,476 feet; and it is traversed in about 35 minutes. A large number of barges make use of the communication afforded by this incline between the Marne and the Ourcq Canal.

Caissons with Water for conveying Barges on Inclines. Large barges carrying heavy loads without injury when floating in water, are liable to be strained if raised out of water when loaded, and conveyed in a cradle along an incline. Accordingly, tanks or caissons have been placed on a framework running on wheels, constructed so as to carry the caisson in a horizontal position up or down the incline, so that barges can be conveyed in them floating in water, and therefore travel on the inclines without experiencing greater strains than in passing along a canal.

Chard Canal Inclines. The above system was introduced about the year 1840, at the Wrantage and Ilminster inclines on the Chard Canal in Somersetshire. These inclines, with a gradient of 1 in 8, had two lines of way upon which two

---

[1] Mémoires de la Société des Ingénieurs Civils, Paris, 1892 (1), p. 627.

caissons, filled with water,' travelled, and counterbalanced each other, being connected by a chain running round a horizontal drum at the top, so that one caisson went up as the other went down; and the motion was imparted by putting more water into the descending caisson. These caissons were 28½ feet long and 6¾ feet wide.

**Monkland Canal Incline.** A similar arrangement, on a larger scale, was adopted, in 1850, for conveying barges up the Blackhill incline on the Monkland Canal near Glasgow, which was constructed to reduce the consumption of water, as the supply was insufficient at times for the passage of the increased traffic through a double flight of eight locks at this place. The incline has a rise of 96 feet, and a gradient of 1 in 10. A double line of way was laid on the incline, with a gauge of 7 feet. A carriage with twenty wheels running on each line of way, was so constructed that it could carry a watertight wrought-iron caisson, 70 feet long, 13⅓ feet wide, and 2¾ feet deep, in a horizontal position on the incline[1]. The two carriages, with their caissons and load of water, counterbalance one another, one ascending as the other descends. The carriages are moved by two engines which turn two vertical drums, in opposite directions, round which the wire rope which hauls the load is coiled. The weight of the carriage, barge, and water, is about 80 tons. When a barge is to be taken up the incline, one of the caissons is immersed in the lower reach, the lower gate of the caisson is raised, the barge is floated in, and the gate lowered. The carriage is then drawn up the incline; and on reaching the top, the caisson is pressed against the entrance channel of the upper reach, which is closed by a lifting gate, so as to form a watertight joint. The gate of the canal, and the upper gate of the caisson are then lifted, and the barge is passed into the upper reach. The whole operation only

---

[1] Minutes of Proceedings Institution C.E., vol. xiii, p. 215 ; and Annales des Ponts et Chaussées, 1877 (1), plate 3.

occupies ten minutes ; and as one barge can be let into the lower caisson, whilst another is being let out of the upper caisson, a barge can be passed up every eight minutes, effecting a saving in time of between twenty and thirty minutes compared with the passage up through the adjacent flight of locks used for descending boats.

Owing to the oscillation produced on the water in the caisson during its motion along the incline, the barge was liable to bump against the ends of the caisson ; and consequently the caisson was only partially filled with water, in order that the barge, instead of floating freely, might touch the bottom and be thus kept still. The barge, accordingly, is only partially supported by the water in its journey along the incline, though sufficiently to prevent the strains produced when a laden barge has to be lifted out of water.

**Georgetown Incline.** A still larger caisson, resting horizontally on a suitable carriage, was constructed for conveying barges of 115 tons on an incline, rising 39 feet with a gradient of 1 in 12, formed at Georgetown on the Potomac, a little above Washington, in 1876, for connecting the Potomac with the Chesapeake and Ohio Canal in place of two locks which had become insufficient for accommodating the traffic. The wrought-iron caisson, 112 feet long, $16\frac{3}{4}$ feet wide, and $7\frac{5}{8}$ feet deep, is supported on three trucks, each having twelve wheels running on four steel rails laid on the incline; and it is drawn up, with its barge, by wire cables worked by a turbine supplied with water from the canal[1]. The caisson, with its load, is counterpoised by four wagons loaded with stone, each provided with sixteen wheels, and running in pairs on a line of way laid with four rails on each side of the incline (Plate 11, Figs. 10 and 11). The total weight of the caisson, with its load of water and floating barge, and trucks, amounted to 390 tons.

---

[1] 'Les Élévateurs et Plans Inclinés pour Canaux.' J. Hirsch, p. 43, and plates 4 and 5.

The incline was worked for a year with the barge immersed in water in the caisson; and the passage between the river and the canal could be effected in about ten minutes. The great weight, however, being concentrated on the three trucks, instead of being distributed all along like the smaller load on the Blackhill incline, damaged the road. Accordingly, the water is withdrawn from the caisson on the entrance of a descending laden barge, to reduce the weight; and the flat-bottomed barge is taken down the incline resting on the floor of the caisson. The emptying and filling of the caisson prolongs the operation; but about 40 barges can be easily passed along the incline in ten hours, whilst the actual transit on the incline occupies only three minutes.

**Ship-Railways.** The railways which have been designed for conveying large vessels across a neck of land, resemble in principle canal inclines on a very extended scale. One of these schemes, namely the Tehuantepec Ship-Railway, has an intimate connection with canals, for it was proposed by the late Captain Eads as a method of connecting the Atlantic and Pacific Oceans, in preference to the construction of a ship-canal at Panama or Nicaragua (Plate 13, Fig 11). Considering that a load of 390 tons, carried by thirty-six wheels on four lines of rails on the Georgetown incline, proved too great for the maintenance of the road, it is evident that a large ocean-going steamer fully laden would require to have its weight distributed over a great number of wheels, running on several very solidly-laid steel rails, to enable it to travel overland without damage to the road. Moreover, as it would be impossible to add to the weight of the vessel, the weight of a caisson with framework to support it horizontally, and enough water to float the vessel, the cradle carrying the vessel would have to be specially designed to prevent the vessel suffering any strain during its passage out of water. Nevertheless, in spite of these practical difficulties in the way of extending the system of canal inclines to the transport of large vessels on railways across isthmuses, the Chignecto Ship-Railway has been constructed, and except for lack of funds would have been completed some time ago, for conveying coasting vessels of 1,000 to 2,000 tons across a neck of land, 15 miles wide, separating the Bay of Fundy from the Gulf of St. Lawrence, and thereby avoiding a stormy detour of between 500 and 600 miles.

## CANAL LIFTS.

Vertical lifts, consisting of two counterbalancing caissons containing water, together with their machinery and guides, serve like inclines to raise and lower barges from one reach of a canal to another, at a considerably different level, in one operation, in a much shorter time, and with far less expenditure of water than the passage through a flight of locks involves. As the caisson, with its load, has to be raised bodily in a lift, instead of being drawn up resting on rails along an incline, more powerful machinery has to be provided for a lift than for an incline with the same load. A lift, however, occupies much less space than an incline, and moreover dispenses with the somewhat cumbrous carriage which is required for placing the caisson horizontally on the incline. The friction, also, of the numerous wheels on which a caisson has to run on an incline, and the maintenance of the lines of way of an incline in perfect order, are avoided in a lift; but, on the other hand, the foundations of a lift have to be exceptionally solid, and its guidance and control perfectly regulated.

**Primitive Canal Lifts.** The first canal lift appears to have been erected in 1809, at Tardebigge on the Worcester and Birmingham Canal, where there are now a long flight of locks and a short tunnel. This lift consisted of a wooden caisson, 72 feet long, 8 feet wide, and 4½ feet deep, supported by iron rods hanging from chains passing over eight overhead cast-iron wheels, 12 feet in diameter, placed in a row on the same horizontal axle ; and the caisson which weighed 64 tons when filled with water, or with a barge floating in it, was counterbalanced by masses of brickwork built on timber platforms, hung by rods from the other ends of the eight chains, and weighing 8 tons each [1]. To maintain the balance exactly when the suspending chains became longer or shorter on one side than the other of the wheels, by the descent or ascent of the caisson or the counterpoises, chains, similar in weight to the suspending chains, were hung from the bottom of the caisson and counterpoises, coiling up on the bottom of the chamber on one side as the suspending chains lengthened, and uncoiling proportionately on the other side as the suspending chains shortened on that side.

The lift of 12 feet was effected in about three minutes, by two men turning the wheels by aid of cogs and pinions worked by winches. Lifting gates at the ends of the caisson and of the reaches of the canal, enabled communication to be opened or closed between the caisson and the canal, the pressure of water on the gates being equalized, before they were lifted, by filling the small space between the adjacent gates of the caisson and canal with water by opening a valve. This primitive lift exhibited the counterbalancing principle and the lifting gates adopted in recent hydraulic canal lifts.

Seven canal lifts very similar in principle, though with smaller caissons, were erected in 1834-6 on the Grand Western Canal between Wellington and Tiverton; but they possessed the improvement of two counterbalancing caissons for the up and down traffic, and were built for lifts having a maximum height of 46 feet. As the small barges of 8 tons navigating the canal were towed in trains, it was important to make the reaches as long as practicable;

---

[1] 'A Description of the patent Perpendicular Lift erected on the Worcester and Birmingham Canal at Tardebig near Bromsgrove.' Edward Smith, Birmingham, 1810.

and, consequently, these high lifts were introduced. The two wooden caissons, strengthened by angle-irons, were suspended at each end of chains passing over three cast-iron wheels, 16 feet in diameter, erected over two vertical chambers, enclosed by arched masonry walls, in which the caissons ascended and descended alternately, the length of the suspending chains being so adjusted that one caisson attained the level of the upper reach of the canal when the other caisson reached the level of the lower reach at the bottom of its chamber [1]. The variations in length of the suspending chains on each side, according to the relative positions of the caissons, were exactly compensated, as in the earlier lift, by suspending similar chains from the bottom of each caisson, whose suspended lengths varied inversely with the lengths of the suspending chains on the same side, and thus preserved the equilibrium between the caissons.

The force required to overcome the inertia of the balanced caissons and the friction of the machinery; was obtained by admitting two inches additional depth of water into the upper caisson, thereby giving it a preponderating weight of one ton. This was effected by arresting the ascending caisson when the level of the water in it was two inches below the water-level of the upper pool of the canal, by means of a forcing bar, so that when the gates were lifted the additional water flowed in. On drawing back the forcing bar, the upper caisson descended, owing to its greater weight, and drew up the lower caisson. The ends of the caissons, and the ends of the upper and lower pools of the canal, adjoining the top and bottom respectively of each chamber, were furnished with lifting gates. When a caisson reached the top or bottom of the chamber, it was pressed tightly by the forcing bar against the adjacent side walls of the canal, between

---

[1] 'The Perpendicular Lifts on the Grand Western Canal.' James Green, Transactions of the Institution C.E., 1838, vol. ii, p. 185, and plates 16 to 18; and Minutes of Proceedings Institution C.E., 1838, vol. i, p. 26.

which the canal stop-gate slided, so as to form a watertight joint. The gates of the caisson and canal were then simultaneously lifted, opening the communication between them. The barges using the lift were 26 feet long and $6\frac{1}{2}$ feet wide, and had a draught of $2\frac{1}{4}$ feet of water; and one barge was taken up the lift, and another let down, in three minutes. The saving in time in the passage of a train of barges from one reach of the canal to the other, was estimated at two-thirds of the time occupied in passing through a flight of locks of similar lift; and the saving in water was 92 per cent.

Besides the counterbalancing of the two caissons, and the lifting gates, these latter lifts, by using a preponderating weight of water in the upper caisson as the moving force, exhibited another feature which has been followed in recent canal lifts. Though, however, these small canal lifts must be regarded as the prototypes of the modern hydraulic canal lifts, and were stated to work satisfactorily, they appear to have soon fallen into disuse; and the system was not extended till, after a lapse of forty years, the first hydraulic canal lift was erected at Anderton, near Northwich, in Cheshire.

**Anderton Hydraulic Canal Lift.** When, owing to the growth of trade on the river Weaver, it became important to connect the river with the Trent and Mersey Canal at Anderton, where the two waterways approach close together, though with a difference of level of $50\frac{1}{3}$ feet, the erection of a flight of locks was first contemplated. The very limited space, however, available for the work, the delay experienced in passing through a flight of locks, and the insufficiency of water in the canal, at the upper level, for supplying the water for lockage, led to the erection of a canal lift, in 1875, instead of a flight of locks, on an island between two channels of the Weaver. Two wrought-iron aqueducts, crossing over the minor river channel, connect the canal with the two caissons of the lift; and a channel was excavated on the island, connecting the lift-pit with the main channel of the river. The lift consists of two wrought-iron caissons, or troughs, each supported under its centre by a cast-iron ram, 3 feet in diameter, moving vertically in a cast-iron hydraulic press, the two presses being connected by a 5-inch pipe, so that the two troughs can be made to counterbalance each other, thereby enabling the descent of one trough to effect the raising of the other [1]. Each trough, 75 feet in length and 15½ feet in width, and containing 5 feet depth of water, can admit one of the largest barges, of 100 tons capacity, navigating the canal, or two of the ordinary barges of 30 to 40 tons. The sides of the troughs are formed by two wrought-iron girders, 9½ feet high in the centre and decreasing to 7½ feet at the ends, which support the weight of the trough and its load of water, or barges and water; and each end is closed by a wrought-iron lifting gate; whilst cross girders, with small longitudinal girders, support the floor. The troughs, each weighing with its load 240 tons, are steadied in their motion by cast-iron guide blocks at each corner, sliding against guides on columns erected at the ends of the lift-pit. Lifting gates, similar to those at the ends of the troughs, weighing 27 cwt. each, close the ends of the two aqueducts connecting the canal with the lift; and each of these gates, being counterbalanced, can be raised by one man in 1½ minutes, so as to afford a headway of 7½ feet above the water-level in the trough or aqueduct, thereby opening communication between the lift and the aqueduct. Watertight joints are formed between the end of the ascending caisson and the end of the corresponding aqueduct, by means of india-rubber strips; and the small space between the two adjacent lifting gates is filled with water, by opening a valve in the aqueduct gate, before opening the gates, as in the earlier lifts. As the lift-pit contains water, being in direct communication with the river by means of the cut, instead of being dry as in the previous lifts, the descending trough has only to be immersed to a depth of five feet in the

---

[1] 'Hydraulic Canal Lift at Anderton.' S. Duer, Minutes of Proceedings Institution C.E., 1876, vol. xlv, p. 110, and plate 2.

water in the lift-pit, and the gate on the river side of the trough to be then raised, to open the waterway between the lift and the river.

The lift is worked by removing six inches depth of water from the trough at the bottom, by means of self-acting siphons, which gives the trough at the top, with its 5 feet depth of water, a preponderating weight of 15 tons; and, consequently, as soon as the communication between the presses is opened, the heavier trough descends, causing the ascent of the other. The descending trough, however, loses its preponderance of weight on becoming partially immersed in the water in the lift-pit, when the ascending trough is about 4½ feet below the top of the lift. Accordingly, when the troughs have reached these positions, the communication between the presses is closed; the press of the upper trough is connected with a hydraulic accumulator; and the final lift of 4 feet, or less than one-twelfth of the whole lift, is effected by hydraulic power stored up by a steam-engine. The ascending trough, with its 4½ feet depth of water, is stopped when its water-level is six inches below the water-level of the aqueduct, in order that the six inches depth of water removed in the lift-pit, may be restored when communication is opened with the aqueducts. The main portion of the lifting to a height of 50⅓ feet is, consequently, accomplished by the consumption of a layer of only six inches of water over the area of the trough for each operation. The lifting can be effected in two and a half minutes; and eight minutes suffice for the operation of transferring two of the smaller barges from the river to the canal, and two others from the canal to the river. The time occupied at Runcorn for a barge to pass through a flight of locks, where the difference in level is the same as at Anderton, is from one hour and a quarter to one hour and a half, showing that the lift effects a great saving of time, as well as of water, when compared with a flight of locks.

The cost of the lift was £29,463, and of the foundations, extending to a depth of 70 feet in a water-bearing stratum, was £18,965, giving a total cost of £48,428. The lift, however, was novel in character, and involved various subsidiary works; and the foundations were made exceptionally heavy, being in the neighbourhood of salt mines where the ground is exposed to subsidence.

After the lift had worked quite satisfactorily during seven years, one of the presses suddenly burst in 1882: and its trough, which was at the top of the lift with a barge in it, fell to the bottom; but as the trough was checked in its descent by the water in the press having to escape through the narrow space of one inch between the ram and the press, and its fall was broken at the bottom by the cushion of water in the lift-pit, very little damage was done either to the lift or to the barge in the trough[1]. Both presses were replaced by thicker cast-iron presses, with slight modifications in form.

**Fontinettes Hydraulic Canal Lift.** The Neuffossé Canal, forming a link of the waterways connecting the North Sea ports of France with Paris and the northern coal-fields, has a large and increasing traffic, so that even by 1874 the passage of a flight of five locks at Fontinettes near Saint Omer, with a total lift of 43 feet, occasioned serious delays, which the subsequent increased draught of the vessels resulting from the deepening of the canal tended to aggravate. The scheme of creating a second flight of locks, approved in 1875, had to be abandoned in 1879, when the increased standard dimensions decreed for the principal waterways of France rendered the existing flight of locks inadequate in size for the proposed navigation. Eventually, in 1881, the erection of a hydraulic lift at Fontinettes was determined upon, in preference to the construction of two flights of locks of the standard type designed to pass vessels of 300

---

[1] Minutes of Proceedings Institution C.E., vol. xcvi, p. 223.

tons through in 30 to 40 minutes, and estimated to cost
£58,400[1].

The Fontinettes lift, which was begun at the end of 1883,
and commenced working regularly in April 1888, is similar
in principle to the Anderton lift, with two balancing troughs
each supported by a central hydraulic ram, mainly worked
by a surcharge of water in the descending trough. The
Fontinettes lift, however, is much larger, having to accommodate vessels of 300 tons; and, as in the earlier lifts,
the troughs descend into a dry lift-pit. The troughs are
129½ feet long, 18⅜ feet wide, and contain 6 feet 6¾ inches
depth of water; and the cast-iron rams are 6 feet 6¾ inches
in diameter, working in presses made of weldless steel coils,
with internal copper lining[2]. The troughs are kept in
position during their motion by central steel guides, embracing
vertical flanges of cast-iron projecting from the main central
building between the troughs containing the machinery,
and from two side towers erected alongside the middle of
the outer sides of the troughs. Moreover, to obviate any
tendency of the troughs to swing round, guides have been
fastened at each side of the up-stream ends of the troughs,
which slide against cast-iron plates fixed into the sides
of two recesses in the masonry pier supporting an aqueduct
which crosses over an adjacent railway, and connects the
lift with the upper reach of the canal. The adjoining pairs
of counterbalanced lifting gates which close the ends of the
troughs and the canal reaches, are locked together before
they are raised by chains suspended from overhead frames
erected at the extremities of the lower reach and aqueduct,
so that no frames have to be put over the ends of the troughs.

The weight of the trough, water, and ram, which has

[1] 'Les Moyens de franchir les Chutes des Canaux.' H. Gruson and L. A. Barbet,
p. 57, Paris, 1890.

[2] 'Canal and River Works in France, Belgium, and Germany.' L. F. Vernon-
Harcourt, Minutes of Proceedings Institution C.E., 1889, vol. xcvi, p. 182.

to be lifted 43 feet 1 inch, amounts to 785 tons; and the working of the lift is effected by introducing a surcharge of water into the top trough. Besides this surcharge of water in the trough, there is also an excess of water in the press of this trough equivalent to the stroke of the ram, amounting to 41 tons, which, however, gradually passes into the other press as the trough descends till, when it reaches the bottom, this surcharge has all passed into the press of the ascending trough. Accordingly, a surcharge of at least 41 tons of water would be needed in the descending trough to maintain its diminishing preponderance till it reaches the bottom of the lift; and it was found in practice that 50 tons surcharge of water were required to complete the operation, which was accomplished in less than three minutes with the communication between the presses fully open. The surcharge, however, introduced was liable to vary with fluctuations in the water-level of the upper or lower reach; and with too great a surcharge, the trough might descend too rapidly, or with too little, it might stop before reaching the bottom. Consequently, the valves regulating the communication between the presses have been partially closed; and the lift is effected in four minutes, by the introduction of a layer of one foot of water into the top trough, weighing 63½ tons. The top trough is lowered a foot to receive its surcharge of water; and it comes to rest on its supports at the bottom when its water level is one foot above the water-level of the lower reach, in order to enable its extra load of water to flow out on raising the gates. The motion of the troughs in working the lift is rapid at first, and gradually slackens to almost nothing at the end; it is controlled by a man stationed in a look-out cabin at the top of the central tower; and interlocking apparatus prevents the troughs being set free to move till the gates are closed and the other preliminary operations completed.

Though the working of the lift is practically effected by the extra weight of water admitted into the top trough, water under pressure is provided with the aid of an accumulator, which, besides serving to raise the gates and turn the capstans, can assist in working the lift when required. This hydraulic power is, moreover, used for lifting one of the troughs from the bottom to the top, when both troughs are in the lift-pit at the commencement of the day's work; and it also raises the upper trough to its position, when it has dropped somewhat owing to leakage of water from its press during a prolonged interval of rest. A turbine of 50 HP, turned by a stream of water from the upper reach, works the pumps which supply the accumulator; and a turbine of 15 HP works an air compressor, for blowing out the air bags which form a watertight joint between the ends of the troughs and the extremities of the canal and aqueduct, and also serves for pumping water out of the lift-pit. The total amount of water expended in each operation is 15,300 cubic feet, out of which only 2,260 cubic feet consist of the surcharge of water, almost the whole of the remainder being expended in working the turbines.

The total expenditure on the lift amounted to £74,960; but this includes the exceptionally heavy cost of £6,600 for land and buildings, and also comprises the cost of working the lift till it was finally handed over; and therefore the actual cost of the lift was under £68,000. Moreover, the contract was let when prices were very high, so that it has been estimated that, under ordinary conditions, a similar lift might be erected at a cost of between £50,000 and £60,000. The time occupied by a vessel in passing from one reach of the canal to the other, varies between 19 and 12 minutes, according to the size of the vessel; but with the average time of 16 minutes, forty-five barges might be passed each way in a day of 12 hours. Notwithstanding numerous stoppages at first for trials, final works, and

modifications, 32,462 barges passed through the lift in the
first 3⅔ years [1].

After the lift had been in successful operation for between
five and six years, one of the presses settled a little in the
sandy alluvial soil, under the influence of the repeated shocks
imparted to the press in working the lift; and this has
necessitated the reconstruction and extension of the foundations, the excavations for which have been carried through
the soft water-bearing stratum by aid of the congelation of
the soil. The locks of the old flight were, in the meantime,
lengthened during the annual stoppage of the traffic, in order
to enable the larger vessels to pass through the flight of locks
whilst the repairs of the lift were in progress.

**La Louvière Hydraulic Canal Lift.** The Canal du Centre
is being constructed in Belgium to connect the Condé Canal at
Mons with the Charleroi and Brussels Canal at La Louvière, so
as to place Mons and its waterways in direct communication
by water with Brussels, Charleroi, and Liége, and thus form
an important link in the midst of those flourishing coal and
iron districts. This new canal, though only 13 miles long,
has to rise 293 feet between Mons and La Louvière, 217 feet
of which have to be surmounted in the 4⅛ miles between
Thieu and La Louvière; and this rapid rise, together with
a scarcity of water, rendered locks unsuitable for the large
traffic that might be expected to use the canal. Accordingly,
it was determined to erect four lifts, like the one previously
designed for Fontinettes, for surmounting the difference of level
in this last section of the canal, with a small expenditure of
water. The Louvière lift, with a rise of 50½ feet, was undertaken as a test of the system in 1885, and was completed
in 1888; but the other three lifts, each having a rise of 55½
feet, have not yet been finished.

The lift at La Louvière is precisely similar in principle to

---

[1] 'Ascenseur Hydraulique des Fontinettes.' Guide-Programme Officiel, Vᵐᵉ
Congrès International de Navigation intérieure, Paris, 1892, p. 51

the Fontinettes lift; but being designed to accommodate vessels up to 400 tons, its troughs have been made somewhat larger, though the cast-iron rams supporting them are the same in diameter (6 feet 6¾ inches) as the Fontinettes rams (Plate 11, Fig. 12) [1]. The troughs are 141 feet long and 19 feet wide; the normal depth of water contained in them is 7 feet 10½ inches, to admit vessels with a draught of 7⅕ feet; and they are borne by lattice girders, in place of the plate girders used at Fontinettes. The presses are made of cast-iron hooped round with continuous weldless steel coils; the length of each press is 64¼ feet, and the length of the stroke 50½ feet; and the working pressure is 469 lb. per square inch. The troughs are kept in position by guides sliding against light wrought-iron braced towers at the centre and at each corner. These towers are connected together at the top by lattice girders carrying a footway round the top of the lift, to which access is obtained by spiral staircases in the towers; and a cabin has been erected over the centre of the lift, from which a man can overlook and control the working by means of levers provided with interlocking apparatus. A wrought-iron aqueduct, crossing over an adjacent high-road, connects the upper reach of the canal with the lift. The troughs descend into a dry lift-pit; and the lifting gates, closing the ends of the troughs and the extremities of the canal reach below, and the aqueduct above, are counterbalanced and raised in pairs as at Fontinettes.

The weight of each trough, with its water and ram, amounts to 1,037 tons; and the lift is worked by admitting 10 inches additional depth of water into the top trough, equivalent to a surcharge of 62 tons, which aided at first by the 48½ tons extra weight of water in the press of the top trough, causes the descent of this trough, and the ascent of

---

[1] 'Canal and River Works in France, Belgium, and Germany.' L. F. Vernon-Harcourt, Minutes of Proceedings Institution C.E., 1889, vol. xcvi, pp. 184 and 202, and plate 6.

the other, in between two and three minutes. The transfer of one vessel of 400 tons from the lower reach of the canal to the upper reach, and of another vessel in the opposite direction, is accomplished in 15 minutes on the average. Two turbines turned by a stream of water from the upper reach, work the pumps for supplying the water pressure through the intervention of an accumulator. The gates are raised, the capstans turned, and the lift-pit kept dry, by means of hydraulic machinery; but the lift can generally be worked by the surcharge of water alone, without the aid of water-power in the presses. Under these conditions, the average expenditure of water is only 7,224 cubic feet for each operation of the lift.

The cost of the lift was £56,200, including £450 for purchase of land. As the three other lifts have not yet been finished, the Louvière lift has not hitherto been able to be used for traffic; but the trials of its working, on its completion in 1888, proved quite satisfactory. Recently, however, it appears that the wall at the upper end of the lift has shifted to some extent, which will have to be put to rights before the lift could be opened for traffic. Nevertheless, this slight failure, and the settlement of the foundation of the press at Fontinettes, have not led to any change in the decision as to the erection of the three remaining lifts on the Canal du Centre, though doubtless special care will in consequence be devoted to the foundations of these lifts.

**Concluding Remarks on Canal Inclines and Lifts.** The value of both inclines and lifts in saving time and water, as compared with a flight of locks, has been conclusively proved. The adoption, moreover, of the counterbalancing principle in both cases has rendered their working comparatively easy; and the relatively small amount of water required for effecting their motion shows how wasteful the expenditure of water-power is in a flight of locks. Inclines are suitable for any amount of rise; they are not dependent on the same delicate adjustments of machinery as lifts ; they are exposed to less risks in case of failure when controlled by powerful brakes ; and they are less costly in construction.  On the other hand, inclines occupy much more space than lifts, which becomes an important consideration where the available space is very limited, as was the case at Anderton, or where land is very costly. Moreover, inclines are not well adapted for conveying barges floating freely in water, on account of the cumbrous, heavy carriage necessitated for supporting a caisson horizontally on an incline, and owing to the forward motion and variable speed on an incline, making a floating barge bump against the sides of its caisson.  The form of the carriage might, indeed, be improved by making the barge travel sideways on the incline, as proposed in a scheme designed for an incline at the Fives-Lille works[1], instead of end on, as hitherto arranged ; but the weight would have to be distributed over a considerably larger number of wheels than on the Georgetown incline, to enable large barges to travel over inclines without injury to the road.

The slight accidents which may disable a hydraulic canal lift, are undoubtedly an objection which may be urged against the extension of the system ; and this, combined with the unsatisfactory working of the Georgetown incline, has led some engineers to propose reverting exclusively to locks with increased lifts, as carried out to some extent on the Canal du Centre, and more particularly as accomplished at the Villette lock on the St. Denis Canal.

Nevertheless, at the present time, hydraulic canal lifts, in spite of slight failures, have been more fully perfected than locks of large lift or inclines, enabling a vessel of 400 tons to be raised in a single lift of 50½ feet, and another vessel lowered the same distance in 2½ minutes, with a comparatively small expenditure of water.  The inclines of the Morris and Monk-

---

[1] 'Plan Incliné pour Bateaux de Navigation intérieure.' A. Flamant, 'Le Génie Civil.' December, 1890.

land canals have, indeed, greater rises; but the barges conveyed along them are quite small in comparison ; whilst the
barges even on the Georgetown incline only attain 115 tons.

Hydraulic lifts, up to the limits of size already attained,
have proved, even in their somewhat trial stage, quite
a satisfactory method for transferring large barges between
canal reaches having a difference of level of 50 feet, with safety,
rapidity, and economy of water; and with still greater care
bestowed upon their foundations, they should be as free from
any risk of a breakdown, as inclines or flights of locks ; whilst
their capacity for traffic is greater than that of these other
systems. The limit of size of lift which can be conveniently
worked by a single ram has, however, been probably reached
at La Louvière; and the extension of the system to greater
troughs, for accommodating a larger class of vessels, will
depend upon the possibility of satisfactorily providing for the
simultaneous action of two or more rams in lifting a single
trough, which was proposed in one of the schemes for the Fontinettes lift. This arrangement, however, has not hitherto been
attempted for canal lifts, though resorted to in the lifting
graving dock at the Victoria Docks, London, for a lift of 25
feet[1], where there is no danger of jamming against guides, and
where exact precision in position is not of the same importance. The great depth required for the foundations of
the presses, and the increase of this depth with any addition
to the height of the lift, appears to preclude the economical
application of canal lifts to much greater heights than those
already reached. The cost of the system is considerable ;
but the difference in cost between a lift and a flight of locks
would probably be more than compensated for in most cases
by the ease of working of the lift, its much greater capacity for
traffic than a flight of locks, and its much smaller expenditure
of water.

---

[1] Minutes of Proceedings Institution C.E., vol. xxv, p. 292 ; and 'Harbours and
Docks,' p. 461.

Inclines possess the important advantage of being equally suitable for large rises as for small ones ; and their cost is merely increased in proportion to their length ; whilst, provided the weight is properly distributed on a sufficient number of solidly laid lines of way, they are as available for large vessels as for small ones. The great additional weight involved in conveying a large vessel floating in water on an incline, would preclude this arrangement being used for ship-railways. Inclines have, hitherto, been only used for much smaller vessels than those accommodated by locks and hydraulic lifts ; but provided arrangements are made for the due distribution of the load on a number of wheels, and a cradle can be designed to prevent a laden vessel being strained in its conveyance overland, the system might be employed for large barges, and is capable of being extended to the carriage of large vessels on ship-railways.

Canal lifts, in their present stage, appear destined to render important services to inland navigation, in facilitating the extension of canals through somewhat rugged districts, by the economy they offer in space occupied, time, water, and working expenses ; but the practicability of their economical application to the accommodation of much larger vessels, still remains to be tested. Locks with moderate lifts have already been employed for ocean-going vessels in ship-canals ; whilst the Villette lock, with a lift of $32\frac{1}{2}$ feet, accommodates the large vessels navigating the Lower Seine ; and there appears to be no serious obstacle to their construction for the largest class vessels, with somewhat greater lifts, as proposed in some of the schemes for the inter-oceanic ship-canals. Inclines also, though hitherto only traversed by small barges, and not much used recently for the extension of inland navigation, have become rivals, in the case of the unfinished Chignecto Ship Railway, of ship-canals for enabling ocean-going vessels to traverse isthmuses. Whilst, however, ship-railways can be more cheaply constructed than ship-canals through rugged country,

owing to the great reduction in the excavation by the substitution of gradients for level reaches, the cost of working ship-
railways, involving the provision of traction overland, and the
maintenance of lines of way exposed to severe wear and tear,
has still to be ascertained by actual practice. Locks,
accordingly, at the present time, afford the only perfectly
assured means of transferring ocean-going vessels from one
reach of a ship-canal to another ; and experience alone can
decide whether inclines, in the form of ship-railways, are
destined to prove formidable rivals of the older method of
canals, for the connection of oceans across isthmuses.

# CHAPTER XVIII.

## IRRIGATION WORKS.

Objects of Irrigation. Sources of supply of Water. Economical aspects of Irrigation Works. Hydrology essential to Irrigation; Data required. Duty of Water for Irrigation. Tanks, in India, construction, loss of water, silting up. Wells, primitive methods of raising Water; supply limited. Reservoirs: instances in India, with heights of dams and capacities; Dams in Spain; proposed Assouan Reservoir; instances in South Africa and North America. Remarks on Irrigation Reservoirs. *Irrigation Canals:* four Classes; Canals from Reservoirs, instances. Inundation Canals, for conveying Water and Silt to the Irrigable Lands, facilitated by straight course and level of River-Bed; choice of Site for Head; measures for protection against Erosion of Banks and Silting near Head. Inundation Canals in Upper Egypt, importance; extent of Irrigation dependent on rise of Nile; Basin System. Inundation Canals in Punjab and Sind, Dimensions, Discharge during Floods; Difficulties occasioned by Silting and Changes in Channel of Indus. Remarks on Inundation Canals.

IRRIGATION consists in supplying water to land to increase its productiveness, and it is, therefore, mainly resorted to in countries where the rainfall is deficient, or occurs only during brief periods, and especially where the dryness of the climate is accompanied by a high temperature, as in the lower latitudes. In some districts, no crops could be raised without a regular supply of water; and in other places water is required to promote the fertility of the soil, to grow certain crops, and to avert periodical famines resulting from a scarcity of rainfall in some years. Many regions of the globe, indeed, need irrigation, either to render them capable of cultivation, or to enhance largely their agricultural value, as for instance Egypt, extensive portions of India, Australia, and the Cape Colony, the arid belt of the western portion of North America, stretching from north to south, and the southern parts of Europe.

**Sources of Supplies of Water for Irrigation.** Water for
irrigation may be obtained, like water for the supply of towns,
from tanks in which the rainfall is collected, from wells by
pumping, from reservoirs formed by dams across river
valleys, or direct from rivers by canals. The supply, however, for irrigation must be more abundant and cheaper, to
meet the requirements of agriculture, than the water-supply
of towns; but the purity of the source of supply is immaterial
for irrigation, and the purification of the water previously to its
delivery is unnecessary. Irrigation works, accordingly, have
to be carried out on a much larger scale, and with more regard
to economy, than water-works for towns.

**Economical Considerations relating to Irrigation Works.**
Irrigation works may be divided broadly into two classes,
namely, those which are essential for the cultivation of the
land they supply, and those which protect districts from
occasional droughts resulting in the failure of the crops and
famine, such as some populous parts of India are exposed to
in specially dry years. The first class of works, if properly
designed and efficiently carried out, in suitable localities,
generally yield an ample return on the expenditure; for the
supply of water provided is certain to be used regularly, as
the land is incapable of cultivation without it; and the
productiveness of well-irrigated lands is usually sufficient to
enable the proprietors to pay easily an adequate rate for the
water on which they are dependent. When, however, irrigation
is only called into use to supply the deficiency of rainfall in years of drought, which occur at uncertain intervals,
the profits obtained in the dry years are very liable not to
suffice to compensate for the absence of profit during the
intervening years, and so fail to yield a proper return on the
capital expended. These protective works should, therefore,
evidently be undertaken by the Government, for the State
alone can reap the indirect profits resulting from a prosperous
condition of the community; and upon the State also must

fall the burden of providing for a famished population, and the loss of the land-tax. The permanent productive irrigation works may, also, be advantageously carried out by the Government, for the State is more likely to extend the works to less productive districts than a private company; it can raise the required capital for the works on lower terms; it can better afford to wait during the period which must elapse before the gradual development of irrigation, and more thorough utilization of the supply of water, can secure an adequate return; and the community is thereby secured from the possible imposition of excessive rates. Moreover, the Government, by taking in hand both classes of irrigation works, can exercise a more thorough control on the working and extension of the system of irrigation throughout the country, with a view to the general benefit of the inhabitants, and can recoup itself for the deficiency in profits on the works for protection against famine, by the large returns from the most successful works. Thus, in British India, the annual return on the whole outlay on irrigation works, amounts to between 6 and 7 per cent.; whilst the interest on the capital expended in the several works ranges from a minus quantity in some cases, up to a maximum of 31 per cent. in the case of the Cauvery deltaic irrigation works.

**Hydrology in relation to Irrigation.** An exact knowledge of the hydrology of a district forms the basis for the design of irrigation works. The nature of the works, and their prospects of success, depend upon the period, amount, and duration of the rainfall, the existence and accessibility of subterranean supplies of water, the flow of the rivers of the locality, and the periods of their floods. The rainfall, indeed, has a twofold influence on irrigation works; for whereas the amount of storage that can be effected, either in tanks or reservoirs, depends on the volume of rain over a given area, a sufficiency of rain in ordinary years renders a district independent of irrigation except in times of drought, which makes irrigation works under such conditions merely protective against famines, and generally unremunerative.

Water-bearing strata, at a moderate depth below the surface, enable wells sunk in them to furnish a supply of water for irrigation, which is raised and used on the adjacent land. These natural underground reservoirs of rainfall, being protected by the overlying stratum from loss by evaporation, provide a simple and frequently-used source of water for irrigation. Artesian wells, also, carried by borings to great depths, when suitably situated and piercing a water-bearing stratum with its outcrop at a considerably higher level, furnish a supply of water rising in the wells.

Irrigation canals drawing their water from large rivers, afford the most abundant and cheapest supplies for irrigation ; and where the general slope of the land is suitable, these canals are able to convey the river water to irrigate districts considerable distances away from the place where the supply is drawn off. The flow, however, and varying levels of the river, require to be accurately known, in order to determine the period of supply, and the maximum volume of water that can with certainty be drawn from the river, and also to make provision for shutting off the floods. passing down the river, that might injure the canal.

**Duty of Water for Irrigation.** The amount of water required for irrigating satisfactorily a given area of land, which is known as the duty of water, constitutes a very important matter in relation to irrigation works, just as the consumption of water per day per head of population does in a town supply. This quantity necessarily depends upon the climate, the nature of the soil and of the crops, and the extent to which irrigation is supplemented by rainfall. The duty of water, moreover, varies somewhat according to the manner in which the irrigation is effected ; and more water is generally required at the commencement of irrigation, than when it has been in operation for some years, owing to the gradual saturation of the subsoil and the raising of the water-line. The volume of water distributed in a year over the land, in various parts of the world, and for different crops, varies from a few inches in depth, up to over a hundred inches in extreme cases where the rainfall is very small; but it ranges generally between about 10 and 60 inches [1]. A sandy soil requires two or three times more water than a clay soil; and, in India for instance, cereals require less water than indigo, indigo less water than rice, and rice less water than sugar-cane.

**Tanks for Irrigation.** In remote periods, tanks were constructed in very large numbers by the natives of India, for collecting the rainfall to provide water for irrigation in the dry season, in places where the rains are abundant but only last for a short time. These tanks are common in Bengal; whilst in Madras, there are about fifty-three thousand tanks of various sizes, enclosed by earthen embankments having a total length of 30,000 miles; and in Mysore, they are still more plentiful, numbering about thirty-seven thousand in a much smaller area. Some of these tanks are only a few acres in extent; whereas the old Veeranum tank, enclosed by an embankment 12 miles in length, has an area of 15 square miles. Many of these tanks are formed by embankments round natural depressions in the land, and depend entirely for their supply upon local rainfall over a small catchment area; whilst others are constructed in stages down a valley, by a series of earthen embankments placed at intervals across the valley, thus resembling reservoirs on a small scale. Vast tracts of rice fields in Madras are irrigated by water drawn from tanks; and more than half the area of Mysore is dependent on this system of irrigation.

The water for irrigation is drawn off through a sluice built in masonry, with its sill level with the bottom of the tank, other outlets being occasionally constructed at higher levels

---

[1] Minutes of Proceedings Institution C. E., vol. lxxiii, p. 210.

to enable lands to be irrigated above the level of the bottom
of the tank. The indispensable escape of surplus water is
generally provided for by a side channel away from the dam,
or occasionally over a waste weir in a portion of the dam
built in masonry, to secure the earthen embankment from being
overtopped by the rising of the water in the tank, for a flow
of water over the bank of earth would soon form a breach.

These open shallow tanks are exposed to considerable
loss from evaporation in hot, dry weather, which may reach a
maximum of nearly half an inch in a day in India at the
hottest period. Absorption and leakage also occasion a loss
of water; but absorption is gradually reduced by the layer of
silt which is deposited over the bottom if the water introduces
sediment into the tank. The capacity of the tank is liable to
be seriously reduced by silting when much deposit comes in
with the water; and this can only be remedied by stirring up
the mud on the advent of the first flood, and washing it out
through the sluice, or by widening and raising the embankments, which involve also the raising of the waste weir and
sluice. Sometimes, in preference to resorting to one of these
expedients, the sluice of a silted-up tank is left open, and crops
are grown in the fertile, silty bed of the tank, affording, in some
cases, a better return than if the water had been stored for
irrigation.

**Wells for Irrigation.** Wells have been a means employed
for obtaining water from remote antiquity; and they have
been extensively resorted to in India for irrigation. Simple
mechanical methods of raising the water from the wells have
been long practised by the natives. The simplest of these
contrivances, in which manual labour alone is employed, is the
*picottah*, which consists of a balanced pole supported and
turning on a high prop, with a rope at one end from which a
bucket is suspended, and a counterpoise at the other end. A
man pulling on the rope raises the counterpoise, and causes the
bucket to descend into the well, which when filled with water

is readily raised again by the help of the descent of the counterpoise. This plan is much used in Bengal for lifts of 4 to 10 feet; and it is also largely employed on the banks of the Nile, where the contrivance is known as the *shadouf.* In the North-West Provinces, Central India, and Bombay, larger lifts are accomplished by aid of the *mote*, in which bullocks draw along a rope in going down an inclined plane, and thus, by means of a pulley hung over the well, haul up a large bucket of water from the well. In Sind, the Punjab, and Egypt, water is raised by a Persian wheel, carrying a continuous chain of pots, from depths of as much as 60 feet, the wheel being turned by oxen or other animal power.

The system of wells thus utilized as a source of water for irrigation, is only suitable for a small supply irrigating land in the immediate vicinity; and it can only be economically employed where manual labour is very cheap. Wells have, however, rendered great services to irrigation in India, and are still very largely used, being retained even in some districts where water from canals is available, for very often the cultivator resorts to his well for irrigation when he has no other employment for himself and his oxen.

**Reservoirs for Irrigation.** In recent times, increased experience in the construction of earthen dams, and the establishment of reliable profiles for the sections of high masonry dams, have enabled reservoirs to be formed by the construction of high dams across the valleys of rivers, storing up water to much greater depths than in the old tanks. This arrangement enables a large volume of water to be retained within a comparatively moderate area, in the upper part of a valley, thereby considerably reducing the amount of land occupied by the reservoir in proportion to its capacity, and diminishing the surface exposed to evaporation, as compared with shallow tanks.

**Instances of Irrigation Reservoirs.** As Bombay possesses few rivers with a good flow throughout the year, and its rainfall is mostly small, irrigation in that province is very largely dependent upon storage; and, consequently, several reservoirs have been constructed for this purpose by the Government in valleys in the hilly districts by the erection of earthen or masonry dams. The Ekruk and Ashti reservoirs are the largest of these formed by earthen dams, having capacities of $123\frac{1}{3}$ and $57\frac{1}{2}$ million cubic yards respectively, and covering areas, when full, of 4550, and 2830 acres[1]; but the maximum height of their dams, of $75\frac{3}{4}$ feet and 58 feet respectively, is exceeded by the Waghad earthen dam, 95 feet high; whilst those of Nehr and Mukti are 74, and 65 feet in height. The largest reservoirs, however, in Bombay are retained by masonry dams; for the Mutha and Bhatgarh reservoirs, with masonry dams reaching heights of 98, and 101 feet, have capacities of 182, and 172 million cubic yards respectively, and spread over areas of 3535, and 3584 acres; whilst the Mhasvad reservoir, formed by a masonry dam with a maximum height of 80 feet, has a capacity of nearly 114 million cubic yards, and a water surface of 4014 acres. The longest of these dams is the Ashti dam, having a length of 12,700 feet; the next in length is the Mhasvad dam, 9080 feet long; whilst the Ekruk dam is 6940 feet long.

A still larger reservoir is in course of formation by the erection of a concrete dam across the river Periyar in Madras, with a maximum height of 155 feet, and a length of 1300 feet. This reservoir will have a capacity of $492\frac{2}{3}$ million cubic yards, of which, however, only about half will be able to be drawn off; but with the large rainfall over the area draining into the reservoir, together with the flow of the river, it is estimated that there will be 1111 million cubic yards of water annually available for irrigation. The reservoir when full will extend over an area of about 7700 acres. This supply of water is designed to irrigate 140,000 acres, a considerably larger area than irrigated at present by any reservoir in Bombay.

Several masonry dams have been erected in Spain within the last three centuries, for the purpose of storing up the waters of

---

[1] 'Irrigation Works in India and Egypt,' R. B. Buckley, p. 86.

streams which are liable to become almost dry in the summer, so as to irrigate the lands which are parched for want of water. The Puentes dam built in 1785-91, is 164 feet high; but this dam, and most of the other Spanish dams, diverge considerably from the rational section established by the calculations for the Furens dam [1]. The productiveness of the fertile soil of Spain is greatly augmented by irrigation, so that the extension of reservoirs in the hilly districts, as well as irrigation from the rivers in the plains, would greatly increase the agricultural wealth of the country.

The construction of a very large reservoir in Upper Egypt is under consideration, for storing up the waters of the Nile towards the close of the flood, when the water is fairly clear, by the erection of a masonry dam across the river near the first cataract, in order to supply water for the irrigation of summer crops in Upper Egypt from March to July, and to extend the irrigable area. Numerous openings will be provided in the dam, to allow the Nile flood to pass freely through it, which will be closed by sluice-gates sliding on free rollers when the water has to be retained towards the end of the flood; and a side channel, with a flight of locks, will afford a passage at the dam for the navigation along the river.

In South Africa, a reservoir for irrigation has been formed in the very dry district of Carnarvon known as Van Wyk's Vley, having a capacity of about 208 million cubic yards, and covering an area of 19 square miles.

Within the dry belt of western North America, numerous sites suitable for reservoirs have been surveyed in the hilly districts of California, Colorado, Montana, and New Mexico, for storing up water for irrigating lands which are incapable of cultivation without its aid [2]. By placing these reservoirs mostly at altitudes of 5000 to 10,000 feet, they will be less

[1] Annales des Ponts et Chaussées, 1866 (2), plate 127, figs. 13 to 20.
[2] 'Twelfth Annual Report of the United States Geological Survey, 1890-91, Part II, Irrigation,' pp. 9 to 208.

exposed to loss from evaporation, and will afford an ample fall for the supply channels, though they will be also further off from the lands to be irrigated. Some reservoirs have been already constructed in California, by the erection of earthen and masonry dams. Thus, for example, the Bear Valley dam, built of granite masonry, 62 feet in height, forms a reservoir having a capacity of 58 million cubic yards ; and a masonry dam, 120 feet high, has been designed for forming a reservoir of nearly ten times this capacity a little lower down the valley. The Sweetwater dam, also, near San Diego, of rubble masonry, 90 feet in height, retains a volume of water of about 35 million cubic yards [1]. These dams are made convex up-stream, to increase their stability by giving them the form of an arch. Mountainous districts are generally suitable for the formation of deep reservoirs ; and rocky foundations for the dams are usually obtainable at a moderate depth below the surface in such regions.

**Remarks on Irrigation Reservoirs.** The supply of water to reservoirs depends upon the area draining into them, and the rainfall over that area, minus losses from evaporation and leakage. Reservoirs possess the great advantage of storing up water, most of which would otherwise be lost by passing down in flood-time ; but reservoir dams are expensive works, necessitating very stable watertight foundations, and the solid compact construction of a well-designed section to secure the valley below against the calamities involved in their failure. Where the rainfall is deficient, and rivers are not available, or their discharge is very variable and liable to fail in dry weather, reservoirs furnish almost the only means of obtaining water for irrigation in countries like the United States, in which the cost of labour prevents extensive irrigation from wells being economically feasible, except where a large supply can be raised by steam-pumping. It is probable, therefore, that reservoirs will by degrees be more generally resorted to in hot dry countries,

---

[1] Transactions of the American Society of Civil Engineers, 1888, vol. xix, p. 201.

so as to store up water in the hilly regions of the upper river valleys for irrigating the lands below.

## IRRIGATION CANALS.

Irrigation canals serve to convey water for irrigation from the source of supply to the lands which are to be irrigated ; and they range from conduits, drawing the supply from tanks or small reservoirs, up to large artificial rivers conveying for many miles considerable volumes of water, from rivers with a large discharge, to lands at a distance from their banks. All these canals have to be given a fall for the water to flow along them, in this respect resembling drainage canals; but some of them have been made navigable by the aid of locks, in the same manner that rivers are canalized.

Irrigation canals may be divided into four classes, namely, (1) Canals conveying the water stored in reservoirs; (2) Inundation canals, drawing their supply from rivers during floods ; (3) Perennial canals, taking water regularly from the upper part of large rivers with a constant flow; and (4) Deltaic canals, branching off from the channels of a delta, and irrigating the intervening low-lying lands.

**Canals from Reservoirs.** The supply of water to these canals is regulated by the outlet sluices of the reservoir; and the sectional area of these canals is determined by the available fall, and the maximum volume required to be passed down in a given period, which depends upon the flow into the reservoir and its storage capacity. These canals resemble the open conduits formed for conveying the water-supply for towns from impounding reservoirs; and wherever the fall of the canal imparts such a velocity to the current as to expose the material forming the bed of the canal to scour, the bottom and sides of the channel must be protected by a lining of concrete or pitching, or the current must be checked by weirs.

Where, owing to a large rainfall, the rivers always furnish an

ample supply of water during the rainy season, and reservoirs
are therefore only required to store up some of the surplus
flow to maintain the supply during the dry season, the canals
leading from these reservoirs are formed large enough to discharge a considerable volume of water, and irrigate extensive
tracts of land.   Thus the Mutha Canal in the neighbourhood of
Poona, 99½ miles in length, has a discharging capacity of 1400
cubic feet per second, and can irrigate an area of 66,150 acres
lying between the canal and the right bank of the river.   The
Nira Canal also, drawing a constant supply from the river
Nira to the south of Poona, by means of the Bhatgarh
reservoir and the Vir basin, which furnish the supply when
the river fails between January and June, can discharge 2100
cubic feet per second ;   and this canal, 101 miles long,
together with 110 miles of distributing channels, can irrigate
56,640 acres of land between the left bank of the river and
the canal [1].

**Inundation Canals.**   Rivers during floods carry down alluvial
matter in suspension, and rising above their banks, deposit
some of this sediment on the adjacent lands which they inun-
date.   Provision has, accordingly, been made in arid and
sterile valleys, to extend this influence to greater distances
from the river, by forming canals branching off from the river,
and thus conveying some of the turbid flood-waters towards
the sides of the valley, which, spreading over the land, saturate
it with water, and fertilize it with the layer of mud deposited.
This process is facilitated by two circumstances: in the first
place, as the canals are carried down the valley in a straight
line, and the river has a winding course, the water in the
canals rises sooner above the level of the plains, or can
irrigate land at a higher level, than the water in the river;
and secondly, these sediment-bearing rivers generally raise
their bed and the adjacent lands, by gradual deposit, to a
higher level than the plains further off from the river.

[1] 'Irrigation Works in India and Egypt,' R. B. Buckley, p. 76.

Inundation canals are generally open channels in direct communication with the river, without any head-works to check the flow into them during floods. Sometimes, however, regulating works are placed across the canal, at a sufficient distance from the river to avoid their being washed away by the shifting of the river-bed, in order to control the discharge into the canal during high floods ; but any arrangement of this kind largely increases the deposit of silt in the canal on the river side of the works. The head of the canal should be placed, if possible, at a stable part of the river ; and places where the bank is being eroded, or a sandbank is forming, should be avoided. If the flow of the river is rapid across the head of the canal, deposit is sure to occur in the canal near the junction, owing to the reduction of the velocity of the current on entering the canal ; and the best position for the head of the canal is in a reach of the river where the flow is moderate, and where the main channel is in the middle of the river-bed. The banks at the entrance of the canal should be protected by pitching, which should be continued a little way along the down-stream bank of the canal, to prevent erosion of the banks at the sides of the entrance, and along the canal near its commencement, which would lead to the deflection of the current in the canal and the consequent formation of a tortuous channel.

The bottom of the canal is formed to a higher level than the bed of the river, so that the heavy detritus, rolled along the bottom of the river, may not enter and obstruct the canal. This heavy matter, moreover, composed of shingle and sand, if it could reach the land instead of settling in the canal, would be of no benefit, for it is only the silty mud which fertilizes the ground. Sometimes, in order to reduce the amount of heavy silt reaching the canal, the head of the canal is formed on a minor flood-channel of the river, whose bed, being a partially silted-up old channel of the river, is at a higher level than the river-bed ; and, consequently, as less sediment enters this channel, and the velocity of the current in it is slower, less sediment gets into the canal. When the canal starts from a branch channel separated from the main river by low islands covered with brushwood and long grass, the water entering the canal is less charged with silt, owing to the checking of the current and the arrest of sediment in the passage of the water through the brushwood and grass. Fortunately the finest matter in suspension, which constitutes the main fertilizing element, is that which deposits least readily. Care must always be taken to avoid any variation of velocity in the current along the canal, for any abatement of speed in the flow causes deposit in the canal; and the silt which is valuable on the land, constitutes a serious difficulty when tending to settle in an inundation canal involving heavy charges for maintenance, or gradually blocking up the canal.

**Inundation Canals in Egypt.** Upper Egypt derives its fertility from the inundation canals which irrigate it with the flood-waters of the Nile, for this portion of Northern Africa is a sandy, rocky waste over which practically no rain falls, so that the cultivation of any crops depends upon the waters of the Nile, supplied directly by canals, or raised from wells in the saturated soil, and also on the red muddy deposit with which the river is densely charged in flood-time, brought from the mountains of Abyssinia. The width of the belt of land bordering the Nile that can be thus irrigated and cultivated, depends upon the amount of fall that can be given to the canals from the river, the depth of the canals, and the height to which the flood rises. The floods of the Nile occur with great regularity, the inundation period extending from August to October; but the height attained by the floods varies, a low Nile flood rising on the average about $4\frac{1}{2}$ feet less at Assouan than a high Nile flood, and a mean flood about $1\frac{1}{3}$ feet less; whilst the heights of extremely high and low floods differ by as much as 9 feet. The ordinary variations, though apparently moderate, exercise a very important influence on the extent of the irrigation, and consequently on the crops, owing to the very small fall that can be given to the canals, so that a deficiency in height below a mean flood of two or three feet produces a serious effect on the year's cultivation.

Generally the canals discharge into basins formed by earthen embankments, which they fill with water to a depth of about 3 feet; and this water in settling on the abatement of the flood, leaves a layer of the Nile mud on the saturated soil, in which the seed is sown in November, and the crop reaped in March. These canals have beds varying in width from 13 feet up to 328 feet, and depths at the river of from 1 foot below a low Nile flood, down to 10 feet below the same level; whilst their fall ranges between about 1 in 20,000 and 1 in 33,000. There are one hundred and three basins on the left bank of the Nile, and sixty-two basins on the right bank, together extending over 1,462,400 acres [1].

**Inundation Canals in India.** Large arid tracts in the Punjab and Sind are irrigated by inundation canals, fed by the floods of the Indus, and its tributaries the Sutlej and the Chenab. The Indus commences to rise in April, owing to the melting of the snow on the Himalayas; it reaches its maximum height in August; and then falls rapidly till October. The flood discharge of the Indus at Sukkur is nearly one-third more than that of the Nile at Cairo; and the amount of silt carried down by the flood of the Indus has been estimated at more than double that brought down by the Nile flood. In years when the flood of the Indus rises to a good height, the land is saturated by the inundation and fertilized by the deposit of silt, as in Egypt; but when the flood rise is deficient, the area irrigated is considerably restricted, and the failure of water is only very partially compensated for by lifting the water with Persian wheels, for without water it is impossible to cultivate the land.

[1] 'Egyptian Irrigation,' W. Willocks, p. 43, and plate 12.

The inundation canals in the Punjab have widths of from 6 to 50 feet, and lengths of from 8 to 60 miles; their fall ranges from about 1 in 4000 to 1 in 10,000; and the water generally flows through them in flood-time from 5 to 8 feet deep[1]. The canals in Sind are of larger dimensions, varying from 10 to 300 feet in width, and reaching 10 feet depth of water; and one of them has a flow of nearly 8000 cubic feet per second during an average flood. Much more water, however, has to be raised from the canals in Sind for irrigation than in the Punjab, owing to their being formed at a lower level in the ground.

The frequent changes in the course of the Indus, resulting from the raising of its bed and the overflow from its banks, occasion considerable trouble in maintaining the heads of its inundation canals, for the retreat of the main channel from the canal leads to the silting up of the entrance, which the velocity of the current renders difficult to clear by dredging; whilst the periodical formation of a new head, to connect the canal again with the diverted river, would be very costly in the case of a large canal[2].

**Remarks on Inundation Canals.** The value of inundation canals depends upon the period and duration of the yearly flood of the river from whence they derive their supply, the amount of fertilizing silt which they spread over the land, and the extent of arid land which they can irrigate. They are only suitable for periodical irrigation; and their efficiency fluctuates from year to year, according to the height and continuance of the flood. They are frequently exposed to reduction in section by the deposit of silt, resulting from changes in the velocity of the turbid stream, necessitating yearly maintenance; and they are sometimes liable to the reduction, or loss, of their supply of water by the silting up of their

[1] 'Professional Papers on Indian Engineering,' 3rd Series, vol. iv, p. 109.

[2] 'The River Indus as a Source of Supply for Irrigation Canals in Sind,' C. S. Fahey, Minutes of Proceedings Institution C. E., vol. lxxi, p. 286.

entrance, or changes in the course of the river.   Accordingly, irrigation by this system is somewhat precarious, especially for lands at some distance from the river, and limited in duration.   Nevertheless, inundation canals have rendered very great services to agriculture in Upper Egypt and parts of India, where cultivation would have been impossible without them; their deficiencies are in great measure due to the inevitable variations in meteorological conditions, which affect the raising of crops more or less in every region of the globe; whilst it is sometimes possible to supplement the periodical inundation thus provided, by storing up a portion of the flood discharge to supply water for irrigation during the dry season, as proposed for Upper Egypt by a dam at Assouan.

# CHAPTER XIX.

## PERENNIAL IRRIGATION CANALS.

Perennial Canals, drawing supplies from Rivers with a fair constant Discharge. Two classes of Perennial Canals, Upper and Deltaic. *Upper Perennial Canals:* Works required; considerations affecting Design of Works; Head-Works, Weir across River, Sluices in Weir, Regulator at Head of Canal. Irrigation and Navigation combined, conflicting Interests. Perennial Canals in India: Western, and Eastern Jumna; Agra; Sirhind; Bari Doab; Sone; Ganges, and Lower Ganges. Canals in Northern Italy: supply derived from Tributaries of River Po; Piedmont Canals, extent, Caluso, Ivrea, Cigliano, Cavour Canal direct from the Po, description; Lombardy Canals, extent, Naviglio Grande, Muzza. Canals in South of France: supply from River Durance; Bouches du Rhône Canals, Crapponne, Alpines, Marseilles for water-supply and irrigation; Verdon, difficulties, siphons; Vaucluse Canals, St. Julien, Cabedan vieux and neuf, Carpentras. Canals in Spain: commenced by the Moors, Jucar, from Rivers Segura and Douro, Aragon, Henares, Esla. Canals in the United States: recent works; Fresno, and Calloway, in California; Del Norte, and High Line, in Colorado; Arizona; Bear River, in Utah; Idaho; Turlock in California. Remarks on Upper Perennial Canals. Cost of several of the canals described. *Deltaic Perennial Canals:* situation; Head-works; small uniform Fall of Canals; Protection by Embankments from Inundation, disadvantages. Deltaic Canals in India; Orissa, Godavery, Kistna, and Cauvery; Extent, Cost, and Revenue. Canals in Lower Egypt, substituted for Basin Irrigation, Object, Disadvantages; Rosetta and Damietta Barrages, and other Improvements. Remarks on Deltaic Canals. Concluding Remarks on Irrigation.

SOME rivers have a sufficiently large constant flow to provide irrigation canals with a supply of water all the year round, instead of merely whilst they are in flood as in the case of inundation canals. Canals drawing their supplies for irrigation from such rivers are called perennial canals, as they enable irrigation to be carried on during any period of the year; and they render the parched lands of tropical countries much more productive than under the system of intermittent irrigation resulting from inundation canals.

These perennial canals are divisible into two classes, namely, Upper Perennial Canals, which derive their supply from the upper parts of rivers, and irrigate a portion of the basins of these rivers; and Deltaic Perennial Canals, drawing their supply from delta-forming rivers at the head of the delta, for irrigating the flat alluvial lands lying between the delta channels.

## UPPER PERENNIAL CANALS.

**General Character and Objects of Works.** The works for a perennial canal, in addition to the formation of the canal and its distributing branches, consist of a weir across the river for raising its level in front of the entrance to the canal. so as to increase the flow of water down the canal, and also regulating works at the head of the canal to control the influx of water, and to prevent the incursion of floods from the river injuring the canal. Waste weirs have to be provided for the escape of surplus water from the canal, due to a stoppage of irrigation on account of a sudden fall of rain, or to the admission of too large a volume from the river; and this precaution is specially necessary where the canal is formed on an embankment, and therefore liable to a serious breach if the water rises above the banks of the canal. Sometimes escapes are formed for the removal of silt from the canal by scour, and also near places where torrents flow into the canal, to provide for the efflux of the surplus water thus admitted. Regulators, moreover, are placed at suitable points for controlling the flow into the branch canals.

**Considerations affecting Design of Works.** The head of a perennial canal, like that of an inundation canal, should, if possible, be placed where the flow of the river is uniform, and its course straight. If the river exhibits a tendency to shift its channel above the site of the weir, it must be regulated by training works to prevent its wandering; and training works are sometimes expedient to direct the flow, at a low stage, closer to the head of the canal. The selection of the site for the head of the canal must be guided by the nature of the district adjoining the river, the position of the lands to be irrigated, and the fall obtainable. The canal must be given a sufficient fall to produce a current adequate to prevent the accumulation of weeds and silt, and yet not rapid enough to cause the erosion of the bed, depending on the nature of the strata through which the canal is formed. As in drainage canals or conduits, the section to be given to an irrigation canal is determined by the maximum volume of water to be discharged, and the inclination of the bed; whilst the side slopes must be regulated by the nature of the soil, unless protected by pitching or concrete. When the available gradient is excessive, it can be regulated by introducing vertical falls or rapids at suitable points, as effected on the Ganges Canal.

By starting a canal from the upper part of a river, an ample fall is obtained for irrigating extensive tracts of low-lying land, and the bed of the river is generally stable in those parts; but the flow is less regular, the sediment brought down consists usually of coarse sand and stones, and the works for conveying a canal across the rugged ground of the upper portion of a river valley, are costly. On the contrary, where a canal starts further down a river, the available fall is considerably less, and the head-works more difficult; but the canal passes over easier ground, it has a much less distance to traverse before reaching the lands to be irrigated, the discharge of the river is larger and more regular, and the silt carried in suspension is of a fertilizing nature.

**Head-Works of Perennial Canals.** The weir placed across the river below the head of the canàl, consists of a solid bank blocking up the low-water channel of the river, and thus keeping up the water-level at a low stage of the river for the supply of the canal; whilst floods pass over the crest of the weir (Plate 3, Figs. 13 and 14). The construction of these weirs varies according to the nature of the foundations, and the intended permanence of the works, as previously described with reference to weirs in India (pages 111 and 112); whilst timber-work is largely used in America for temporary dams in the summer, and crib-work filled with rubble stone for more permanent weirs. In some cases, the raising of these solid weirs to a height sufficient to maintain the water above them at the desired level during the dry season, might result in the diversion of the river when the floods come down; and then it is necessary to restrict the height of the solid portion, and to place a temporary barrier on the top, which can be lowered in flood-time, such as a row of shutters hinged to the apron (Plate 4, Fig. 13).

As a solid weir across a river, by arresting the travel of detritus along the bed of the river, causes an accumulation of deposit in the river-bed above it, sluices are generally placed in the weir near the bank where the head of the canal is situated, in order to maintain a deep channel in front of the head by the scour through the sluices, so that the supply of water to the canal at a low stage may not be impeded. The openings of these sluices, from 5 to over 20 feet in width, with their floors level with the bottom of the river, are closed by draw-doors, movable shutters, or needles (pages 113, 123, and 134); and the total width given to a set of sluices should be determined by the size of channel required, the amount of silting, and to some extent by the size of the river, the nature of its floods, and the height of the weir. Formerly, in India, sluices were also placed in the centre of the weir, for maintaining the channel of the river; but now a low weir, surmounted by a movable dam, is preferred for this object.

The regulator across the head of the canal consists of a set of sluices, generally closed by draw-doors, but sometimes by movable shutters, rolling-up curtains, planks, or needles. A bridge supported by the piers between which the sluice-gates closing the openings slide, is carried over the sluices, from which the gates are raised and lowered; and the top of the bridge is placed above the highest flood-level of the river, so as to be always accessible, and to prevent the floods flowing over into the canal. The dimensions of the sluice openings are determined by the supply of water to be admitted, and the available head of water in the river. The level of the floor of these sluices is regulated by the nature of the silt brought down by the river; for heavy silt, which is useless for agriculture, is rolled along the bottom, and must be excluded from the canal; whereas light silt in suspension is generally valuable, and should be conveyed to the irrigated lands. The raising of the bottom of the inlet above the river-bed, to keep out the dense sediment, necessitates a proportionate heightening of the weir to secure an adequate flow into the canal in the dry season.

**Irrigation Canals adapted for Navigation.** Sometimes irrigation canals have been designed to serve a double purpose, by the introduction of locks to render them navigable, as for instance in side channels at the falls on the Ganges Canal. This combination proved chiefly successful before water-carriage was exposed to competition by railways, and especially where the canals traverse extensive flat plains, as in Lombardy, and in deltaic regions where no locks are required. At the present day, except in the flat regions referred to, the conflicting interests of irrigation and navigation cannot be economically adjusted. Irrigation canals require a current of at least $1\frac{1}{2}$ to $2\frac{1}{2}$ feet per second to discharge their supply, and to keep them free from weeds and from the deposit of the silt which they carry along to fertilize the land; they should be reduced in section as the distributing canals branch off; and their course is determined by the position of the lands to be irrigated. Navigation canals, on the contrary, should have as little current along them as possible; they should be fed with water free from silt; they must have a uniform section and depth of water throughout; they should be divided into long level reaches; and they should be carried past the main centres of trade, to secure an adequate traffic. Accordingly, in face of railway competition, and owing to the cheapness of carriage along roads in the agricultural districts of India, it appears to be inexpedient to attempt in future to combine irrigation and navigation, except under specially favourable conditions.

**Perennial Canals for Irrigation in India.** The earliest perennial canal constructed in India, outside deltaic regions, appears to have been the Western Jumna Canal in the Punjab, formed in the fourteenth century, and extended in the seventeenth century to Delhi, when the Eastern Jumna Canal was also constructed, both drawing their supplies from the Jumna near the Siwalik hills. Since 1870, these canals have been reconstructed. The Western Jumna Canal has a total length, with its branches, of 280 miles, and 908 miles of distributing canals; it has a maximum discharge of 2800 cubic feet per second, and irrigates over half a million acres of land; and the Eastern Jumna Canal, with a maximum discharge of 2350 cubic feet per second, irrigates 240,000 acres.

The Agra Canal, completed in 1874, also draws its supply of 1500 cubic feet per second from the Jumna, a weir half a mile long having been placed across the river at Okhla, ten miles below Delhi; and it can irrigate 240,000 acres of land on the right bank of the Jumna, between Delhi and Agra. This canal, 137 miles long, has a bottom width of 70 feet, a depth of water of 10 feet, and a gradient of 1 in 10,560.

The Sirhind Canal, constructed in 1869–82, starts from the Sutlej above Roopur, where the river emerges from the Siwalik hills, and irrigates about 800,000 acres of the Punjab lying to the south of the river, having a maximum discharge of 6000 cubic feet per second. The main canal is 41 miles long, from which branch canals diverge, 503 miles in length, with distributing branches having a total length of 4400 miles. The bottom width of the main canal is 200 feet, and the inclination of its bed is 1 in 8000. The works along the first part of this canal, as in the case of all canals commencing near a mountainous district, were heavy, the country being rugged and several torrents having to be crossed.

The Bari Doab Canal, commenced in 1850, derives its supply of water from the river Ravi, a tributary of the Chenab, near Madhopore, where the river descends from the hills; and it irrigates 525,000 acres of land in the Punjab lying between the Ravi and the Beas, a tributary of the Sutlej, to the north-west of the Sirhind Canal system. The maximum discharge of the canal is 4000 cubic feet per second; and the main canal with its branches extends over 362 miles, the water being distributed by 1058 miles of minor branches.

**Sone Canals.** The Sone Canals, constructed in 1869–74, draw their supply from the river Sone, which is backed up at Dehree by a weir, 2⅛ miles long; and high floods rise 8½ feet over the weir (Plate 3, Fig. 13). Movable iron shutters, 18 feet long and 2¼ feet high, are placed along the crest of the weir to increase its height in the dry season; and the side sluices are closed by shutters hinged to the floor, and fitted with hydraulic brakes (p. 135, and Plate 4, Fig. 13). These canals, 370 miles in length, with 1200 miles of distributing branches, irrigate at present about 370,000 acres lying between the Sone and the Ganges above their confluence, though they are intended eventually to irrigate 1,016,000 acres; and their maximum discharge is 6000 cubic feet per second, the bottom width of the main canals being 180 feet, with a gradient of 1 in 10,560. The canals for some distance from their head are much impeded by coarse sand and silt brought down by the river during floods, which are deposited along the first four or five miles of the canal, and are only removed by several dredgers working during two or three months in the flood season [1]. The silt contained in the river water might have been carried forward to the irrigated lands, by giving the canals a greater inclination than 6 inches per mile, which provides a maximum velocity of $2\frac{1}{2}$ feet per second, falling often to under 2 feet. The sand rolled along the bottom of the river has raised the river-bed above the weir up to its crest; and it could only have been excluded from the canal by raising the inlet above the level of the side sluices, or possibly by some modification in the position of the head of the canal and the approach.

**Ganges and Lower Ganges Canals.** The Ganges Canal, constructed in 1848–55 and subsequently improved, is the largest irrigation work of this kind; it starts from the Ganges just below Hurdwar, about twenty miles above Roorkee; and the flow of the river during the dry season is diverted into the minor Hurdwar channel by training works, two weirs, and three temporary dams, from which it is led into the canal by means of a weir across the channel. Within the first few miles of its course, some torrents have to be carried over the canal; and several falls have been introduced. Just before reaching Roorkee, the canal crosses the valley of the river Solani on an aqueduct, 920 feet long, with a maximum height of 24 feet to the bottom of the canal, and fifteen arches of 50 feet span; and the waterway, 172 feet wide, conveying the maximum discharge of the canal of about 6800 cubic feet per second, is divided into two channels by a wall along the centre, so as to enable half the aqueduct to be closed for repairs [2]. The main and branch canals of this system have a total length of 440 miles, with distributing branches about 2500 miles long; and they are capable of irrigating 1,600,000 acres of land lying between the right bank of the Ganges and the Jumna.

The Lower Ganges Canal, designed to supplement and

---

[1] 'Irrigation Works in India and Egypt,' R. B. Buckley, p. 39.
[2] 'Report on the Ganges Canal Works,' Col. Sir Proby T. Cautley, vol. ii. p. 417.

extend the irrigation of this district down to, and below
Cawnpore, was commenced in 1871, and partially opened in
1878; it draws its water from the Ganges at Narora by the
help of a solid masonry weir across the river, 10 feet in height,
surmounted by falling shutters 3 feet high. This canal, with
a bottom width of 266 feet, a depth of water of 10 feet,
and a gradient of 1 in 10,560, has a maximum discharging
capacity of 5100 cubic feet per second, and is intended
eventually to irrigate about 1,190,000 acres, of which only
about one half is irrigable at present. This system comprises
about 560 miles of main and branch canals, and 2100 miles
of distributing branches. The main canal has a bottom width
of 216 feet, a depth of water of 8 feet, and a gradient of
1 in 10,560. The canal is carried across the valley of the
Kali Nadi on an aqueduct of fifteen arches of 60 feet span,
the largest work of the kind in India. The low velocity of
the current in the canal, of about only 2 feet per second,
promotes the deposit of silt and the growth of weeds, the
silt being specially deposited in the first two miles of the
canal; but the sediment is scoured out by means of an escape
into the river lower down.

**Canals for Irrigation in Northern Italy.** The construction
of canals in the valley of the Po, both for irrigation and navigation, was commenced in the twelfth century in Lombardy,
and in the fourteenth century in Piedmont. Like the canals
irrigating the upper basins of the Indus and the Ganges, the
canals of Northern Italy draw their supply from rivers fed in
summer by the melting of the snow upon high mountain
ranges; but whereas the canals of Piedmont are fed by rivers
which are irregular in their flow, coming direct from the Alps,
the canals of Lombardy derive their supply from rivers issuing
from the large lakes of Northern Italy, which act as regulators
of their discharge (p. 17). Till past the middle of the nineteenth century, no attempt was made to utilize the waters of
the Po for irrigation, owing to the difficulties anticipated in

carrying a canal across the basins of the tributaries on the north side of the Po, which flow into the Po approximately at right angles to its course, instead of gradually converging towards the main river like the tributaries of the Indus and the Ganges. The lands irrigated lie between the Alps and the left bank of the Po ; and the water is obtained from the tributaries on this side of the river.

**Irrigation Canals in Piedmont.** Before the construction of the Cavour Canal, the total area irrigated in Piedmont was 486,600 acres, and the maximum supply of water 8290 cubic feet per second; whilst the canals and their branches had a total length of over 1200 miles [1]. The oldest of these canals of any importance is the Caluso Canal, 20 miles long, constructed in 1556–60, which starts from the left bank of the river Orco near Spinetto, and conveying a maximum supply of about 350 cubic feet per second, irrigates 18,000 acres of land. The gradient of the canal is irregular, as the natural slope of the country was followed, and it averages nearly 17 feet per mile ; its bottom width is 18 feet ; and the average depth of water in it is about $3\frac{1}{2}$ feet.

The Ivrea Canal, starting from the river Dora-Baltea at Ivrea, is $44\frac{3}{4}$ miles long, and has a fall of 4 to $5\frac{1}{2}$ feet per mile ; and discharging a maximum of 700 cubic feet per second in the summer, it irrigates 30,000 acres. This canal, opened first in 1468, abandoned in 1564 owing to its having become choked with sand from the Dora-Baltea, and re-opened in 1651, extending with its branches over 100 miles, appears to have been the largest canal in Piedmont previously to the formation of the Cavour Canal. The Cigliano Canal, however, also drawing its supply from the Dora-Baltea, constructed in 1783–90, 10 miles in length, with a maximum discharge of 650 cubic feet per second, and irrigating 32,500 acres, having been widened about seventy years later, from a bottom width of $26\frac{1}{2}$ feet near its head, to double this

---

[1] 'Italian Irrigation,' Capt. R. Baird Smith, vol. i. pp. 111, 122, and 157.

width, has now a discharging capacity of 1760 cubic feet per second [1].

**Cavour Canal.** The Cavour Canal was constructed in 1862–66, with the object of supplementing the existing irrigations when deficient, but especially to extend the irrigation over the lands lying between the rivers Sesia and Ticino, which in some parts depended in a large measure upon the torrential rivers Agogna and Terdoppio. The canal starts from the left bank of the Po near Chivasso, about twelve miles from Turin; and diverging from the river in a north-easterly direction, it terminates in a junction with the Ticino above Novara, after a course of 53 miles. The canal has been given a gradient of nearly 16 inches per mile, and a bottom width of about 65 feet along most of its length. The canal is carried over the Dora-Baltea, at a distance of 7 miles from its head, on an aqueduct, 635 feet long, having nine arches of $52\frac{1}{2}$ feet span, as well as over some minor tributaries; and it is carried under the Sesia, Elvo, Agogna, and Terdoppio in siphon culverts, as this canal was designed solely for irrigation. The supply authorized to be drawn from the Po was 3885 cubic feet per second; but on the completion of the canal, the minimum summer discharge of the Po was found to have been much overestimated, which necessitated the construction of a special canal, about two miles long, to introduce the requisite supplementary supply into the canal eight miles from its head, drawn from the Dora-Baltea which, rising in the Alps, has a sufficient discharge in the summer, from the melting of the snow, to make up the deficiency in the supply of the Po, as well as to furnish water for the older irrigation canals which branch off from it. The level of the water at the head of the canal is raised by a weir across the Po; and the supply is admitted into the canal through twenty-one sluice openings, nearly 5 feet wide and $7\frac{1}{2}$ feet high, which are closed by wooden sluice-gates sliding

---

[1] 'Irrigation in Southern Europe,' Lieut. C. C. Scott Moncrieff, p. 227.

vertically in grooves in the masonry piers, little rollers being fitted on the down-stream sides, resting against the grooves, to reduce the friction. The canal is capable of irrigating about 490,000 acres of land; but though it has proved of great value for irrigation, its great eventual cost, amounting to about £4,100,000, prevented its proving commercially successful to the private company who constructed it; and the concession was only granted for fifty years.

**Irrigation Canals in Lombardy.** The total area irrigated in Lombardy, by the canals drawing their supplies mainly from the Ticino, Adda, and Oglio, amounts to about 1,061,300 acres; and the main lines of canals have a total length of about 113 miles; but with their main branches the country has been estimated to be intersected by over 4500 miles of canals[1]. The supply of water obtained by these canals from the above rivers, and from minor rivers and springs, amounts to a maximum of about 15,000 cubic feet per second. The main lines of canals were for the most part constructed between the twelfth and sixteenth centuries; though the irrigated area was subsequently extended, and the Pavia canal, commenced in the fourteenth century, was only completed in the nineteenth.

The earliest of the canals constructed in Lombardy, and also one of the largest, is the Naviglio Grande, commenced at Tornavento on the Ticino in 1177, or possibly earlier, and extended from Abbiategrasso to Milan in 1257, having a length of 30½ miles. The upper portion of this canal varies in width from 75 to 160 feet, and in depth from 4½ to 15 feet; and its gradients range between 13 inches and 8 feet per mile, so that many parts of the bed of the canal have to be protected from scour by pitching. The water-level at the head of the canal is kept up by a weir across the Ticino, 918 feet long, composed of masonry, concrete, rubble, and timber. The discharge of the canal for

[1] 'Italian Irrigation,' Capt. R. Baird Smith, vol. i. pp. 203, 245, and 296.

irrigation amounts to 1850 cubic feet per second; and the canal serves to irrigate 93,440 acres.

The Muzza Canal, connected with the Adda in 1220, and enlarged towards the end of the century, is the largest irrigation canal in Lombardy. It is 43½ miles long, and has a discharging capacity of 2650 cubic feet per second; and it can irrigate 182,500 acres. As in the case of the Naviglio Grande, the width of the Muzza Canal is very irregular, varying from 80 to 180 feet; and, in spite of several weirs across the canal, having a total drop of 61 feet, the gradient of the Muzza Canal averages 4¾ feet per mile. The flow of the Adda is directed into the canal by an oblique weir of masonry across the river near the head of the canal.

**Irrigation Canals in the South of France.** Portions of the departments of the Bouches du Rhône and Vaucluse, lying between the Durance and the Rhone to the south, and between the Rhone and the foot of the Alps on the north side of the Durance, have been for a long time watered by irrigation canals. The river Durance is peculiarly well suited for supplying irrigation canals; for being fed by the snows from the Alps, it has a good discharge in the hot weather; it flows at a considerable elevation, so that there is an ample fall available for the canals branching off from it; and its waters contain a considerable quantity of fertilizing silt which, even when spread over the gravelly plain of La Crau to the east of the Rhone delta, renders this desert land very productive.

**Bouches du Rhône Irrigation Canals.** The first canal in this district was undertaken by the engineer Crapponne, in 1554, mainly at his own expense; but having encountered great financial difficulties, and infiltrations having occurred through the porous soil, he only obtained a flow of water through the canal to Salon in 1559; and the branch canals, with their distributing channels, were only slowly extended[1]. The Crapponne Canal starts from the left bank of the

[1] Annales des Ponts et Chaussées, 1874 (1), p. 271, and plate 4, fig. 9.

Durance, nearly opposite Cadenet, about 37 miles above the confluence of the Durance with the Rhone; and the water is directed into the canal by an oblique weir constructed of fascines, timber trestles, and boulders; this somewhat temporary form of dam, though necessitating frequent repairs, having been adopted by Crapponne, and subsequently continued, owing to the difficulty and cost of constructing a solid weir across the wide shifting channel of the Durance, encumbered by shoals of gravel and sand. The supply is admitted into the canal through eight sluice openings, nearly 4 feet in width; and the maximum discharge utilized is 500 cubic feet per second, irrigating 24,000 acres of the plain of La Crau, though the canal and its branches have a sufficient capacity to pass a considerably larger volume, and could irrigate a much larger area [1]. The main canal, 26 feet wide and 6½ feet deep, and with a mean velocity of flow of 5 feet per second, extends from the Durance to Lamanon, having a length of 20½ miles, where it branches off to Salon, Istres, and Arles; and the total length of these canals is about 90 miles.

The Alpines Canal and its branches, commenced in 1773, start from the Durance at Mallemort, about 8 miles lower down the Durance; they extend over a length of about 194 miles, and distribute 770 cubic feet per second for the irrigation of 20,000 acres of the plain of La Crau. An oblique weir of timber trestles, fascines, and boulders, directs the flow of the Durance towards the head of the canal, as in the Crapponne Canal and in most of the other irrigation canals drawing their supply from the Durance.

**Marseilles Canal.** The Marseilles Canal, constructed in 1837–48, draws its supply from the Durance opposite Pertuis, 9 miles above the intake of the Crapponne Canal, and 614 feet above sea-level. This canal, with a maximum capacity of flow of 424 cubic feet per second, serves the

[1] 'Irrigation in the South of France,' G. Wilson, Minutes of Proceedings Institution C. E., vol. li. p. 219, and plate 5.

double purpose of supplying Marseilles with water and irrigating the districts for some distance round Marseilles. The main canal has to traverse such a hilly country between the Durance and the neighbourhood of Marseilles, that, though it follows a very circuitous course, it passes through 10 miles of tunnels in a distance of 51½ miles. The ruggedness of the country crossed necessitated numerous other important works, of which the most remarkable is the aqueduct of Roquefavour, carrying the canal over the valley of the river Arc[1]. This aqueduct, built with three tiers of arches having spans of 52½ feet, is 1253 feet long, and has a maximum height of 271 feet; and it has been given a special gradient of 1 in 250.

The main canal has a bottom width of 10 feet, slopes of 1⅓ to 1, and a depth of nearly 8 feet; and its gradient is 1 in 3333, except in tunnels and along aqueducts, where the gradients have been increased to 1 in 1000 and 1 in 1428 respectively, allowing a smaller channel to carry the supply. The canal goes on to Montredon to the south of Marseilles; and the total length of the canal with its branches amounts to 98½ miles. A masonry weir, in this instance, was constructed to retain the water in front of the head of the canal, with its crest slightly above the low-water level of the river, except in the main channel, where it is kept 5 feet lower; but the weir can be temporarily raised when required, by movable shutters along its crest[2].

**Verdon Canal.** The Verdon Canal was constructed in 1863–75, for the purpose of irrigating about 42,000 acres of land in the district round Aix, with a volume of 212 cubic feet of water per second. It starts from the left bank of the river Verdon, a tributary of the Durance, near the

---

[1] 'The Bridge-Aqueduct of Roquefavour on the Canal of Marseilles,' G. Rennie, Minutes of Proceedings Institution C. E., vol. xiv. pp. 190 and 202, and plates and 2.

[2] 'Irrigation in Southern Europe,' Lieut. C. C. Scott Moncrieff, p. 50, and plate 7, fig. 4.

village of Quinson; and after a course of 51 miles, it terminates near Aix [1]. Owing to the hilly country traversed, it passes through 12½ miles of tunnels, and along three aqueducts, the largest of which carries the canal across the Paronvier valley on twelve arches of 26½ feet span, its length being 470 feet and maximum height 70½ feet.

The most peculiar feature in the works of this canal is the adoption of siphons for crossing several of the valleys, instead of aqueducts, on account of the conditions relating to the slope, and for the sake of economy [2]. The siphons were tunnelled through the rock forming the bed of the valley, and lined with masonry. In the valley, however, of La Lauvière, though the side slopes are rock, the bottom consists of clay and fissured rock; and therefore the central portion of the siphon, in this case, was formed by a wrought-iron tube, 7½ feet in diameter like the rest of the siphon, carried across the bottom of the valley on cast-iron saddles borne by rollers resting on masonry supports, to allow for expansion and contraction, and fitted at each end into the portions tunnelled in the side slopes. Vertical shafts at the two extremities, sunk into the rock, connect the canal with the siphon, as in the other siphons; the total horizontal length of the siphon is 894 feet; its maximum dip is 77 feet; and the hydraulic gradient between the ends is 1 in 909. This use of a wrought-iron tube was subsequently extended to the whole of the siphon across the valley of Saint-Paul, 948 feet in width where the canal crosses, and dipping 118 feet below the canal. An attempt was first made to form the siphon in the rock; but a portion of it failed under the pressure of the water where the rock was fissured. Accordingly, two wrought-iron tubes, 5¾ feet in diameter, were laid side by side, 13 feet apart, across the valley, supported, like the previous

---

[1] 'Canal du Verdon,' M. de Tournadre, Annales des Ponts et Chaussées, 1881 (2), p. 15, and plates 16 to 20.

[2] Annales des Ponts et Chaussées, 1876 (2), p. 450, and plates 22 and 23.

tube, about 3 feet above the ground [1]. The tubes on each
side, laid to slopes of 2½, and 2¾ to 1, are 251, and 277
feet long on the upper and lower side respectively; the
central portion, laid horizontally across the bottom of the
valley, is 323½ feet long; and the hydraulic gradient between
the ends of the siphon is 1 in 990. Two tubes were preferred to a single one of 7½ feet in diameter, to ensure the
continuity of the supply in the event of an accident or
repairs to one of the tubes. The cost of the siphon was
£10,175, whereas the estimate for an aqueduct was £16,000;
and, moreover, the building of an aqueduct would have
required two years, in place of the six months in which the
siphon was completed.

The main canal, when in an open trench, is 6¼ feet deep,
has side slopes of 1⅓ to 1, and a depth of water of 5 feet;
and with the average gradient of 1 in 4760, the bottom
width is 11½ feet, but the width is varied according to
the gradient. The gradient has been increased in the
tunnels to 1 in 1000; and the section of the channel has
been made 6 feet 7 inches wide, and the same in depth.
A solid masonry dam had to be erected across the Verdon
just below the head of the canal, in order to raise the water-
level of the river sufficiently to ensure an adequate flow
into the canal [2]. The dam, with its concrete foundation,
has a total height of 59 feet, and its crest is 40 feet above
the low-water level of the river; and the dam has been
arched up-stream to increase its stability. The canal and
its eight branches have a total length of about 98 miles.

**Vaucluse Irrigation Canals.** A canal appears to have
been made in the twelfth century, drawing its supply from
the right bank of the Durance, for working a mill near the
St. Julien gate of Cavaillon, situated close to the river. This

---

[1] 'Siphon métallique de Saint-Paul,' M. Bricka, Annales des Ponts et Chaussées,
1877 (I), p. 372, and plates 4 and 5.

[2] Annales des Ponts et Chaussées, 1872 (I), p. 432, and plates 11 and 12.

canal, known as the St. Julien Canal, was enlarged and utilized in the following century for irrigation, and was gradually extended, so that now it irrigates about 4670 acres[1]. It starts from the Durance about 8 miles above Cavaillon, and running in a direction nearly parallel to the river, it extends about 6 miles below this town; and the supply is turned into the canal by one of the somewhat temporary dams of timber, fascines, and boulders, commonly used for this purpose on the Durance.

Two canals run parallel to the St. Julien Canal between this canal and the foot of the hills, namely, the Cabedan Vieux and the Cabedan Neuf, the one starting from the Durance, a little below the head of the St. Julien Canal which it crosses at a short distance from the river, and the other diverging from the Durance at Merindol, about five miles higher up, constructed in the eighteenth century. These canals have been gradually extended beyond the river Coulon, under the names of the Fugueirolles, Plan Oriental, and L'Isle canals, for irrigating the wider plain opening out into the Rhone valley, the waters of the canals being directed across the torrential Coulon by a dam. The part of the plain nearer the confluence of the Rhone and the Durance is irrigated by several other canals, drawing their supplies from the Durance, and from the river Sorgues which issues from the fountain of Vaucluse amongst the hills.

The largest canal of this district is the Carpentras Canal, deriving its supply from the Durance, in conjunction with the Cabedan Neuf and L'Isle canals; it was commenced in 1854, and skirts the base of the hills bounding the plain on the left bank of the Rhone, passing by the town of Carpentras from which it takes its name. This canal is 32 miles long; its gradient is 1 in 4000, and the rate of flow of its current $1\frac{3}{5}$ to 2 feet per second; it has a width of 33 feet, a depth of water of $2\frac{4}{5}$ feet, and a maximum

[1] Annales des Ponts et Chaussées 1850 (2), p. 338, and plate 194.

discharge of 212 cubic feet per second; and it irrigates 16,800 acres. The canal is carried over the river Coulon by an aqueduct; and another aqueduct, with thirteen arches of 29½ feet span, conveys the canal over the river Sorgues near Vaucluse, at a height of 78 feet above the river.

**Irrigation Canals in Spain.** With a warm climate, a small rainfall, and several important rivers traversing extensive plains, Spain is a country which should abound in irrigation works; but though irrigation canals were carried out by the Moors in Spain, on a scale which may be regarded as extensive, considering the early period at which the works were carried out, comparatively little progress has been effected during the long period which has elapsed since the expulsion of the Moors from the country. Tunnels and siphons adopted by the Moors for carrying their irrigation canals across rough country, display their ability in designing such works.

One of the earliest and largest of the irrigation canals is the Jucar Canal, 25 miles long, deriving its supply from the river Jucar at Antella, amounting to about 900 cubic feet per second, and irrigating about 31,000 acres of land in Valencia. Several smaller canals branching off from the river Guadalaviar, or Turia, irrigate about 26,000 acres in Valencia, the works in this province having been carried out by the Moors in the eighth or ninth century. The river Segura, flowing through the provinces of Murcia and Alicante, supplies numerous canals for the irrigation of about 85,000 acres of the extensive plains in the districts of Cieza, Murcia, Elche, and Orihuela [1]. The rivers Darro and Genil, also, irrigate about 47,000 acres in Granada, by means of canals formed by the Moors. Masonry weirs across these various rivers turn the water into the canals.

The Imperial Canal of Aragon, commenced by the Em-

[1] 'Irrigations du Midi de l'Espagne,' Maurice Aymard, pp. 17, 75, and 199, and plates 2, 3, and 10.

peror Charles V in 1528, but left unfinished, was at last
completed in the latter part of the eighteenth century by
Charles III of Spain, during whose reign several important
irrigation works were carried out or extended [1]. This canal,
drawing its supply of 880 cubic feet per second from the
river Ebro, runs parallel to this river between Tudela and
Saragossa, and extending 3 miles beyond the latter town,
has a length of 56 miles; and it irrigates about 67,000 acres
on the right bank of the Ebro, having cost about £15 per
acre irrigated, though it also serves for navigation.

The most recent important irrigation canals constructed in
Spain are the Henares and Esla canals, carried out in 1863–
68 for irrigating the valleys of the rivers Henares and Esla,
with water drawn from these rivers [2]. The river Henares
is a tributary of the Tagus; and the head of the canal is
situated 16 miles above Guadalajara, where a masonry
weir across the river, 390 feet long and curved up-stream,
directs the water towards the canal, the maximum supply
granted being 175 cubic feet per second. The canal is
28 miles long, and can irrigate 27,000 acres; its gradient
averages 1 in 3067; and the velocity given to the current is
2⅛ feet per second, which suffices to prevent the rapid growth
of weeds in the canal. The canal is carried through a limestone cliff overhanging the river, in a tunnel 1⅘ miles long; and
it crosses over the Majanar torrent in a wrought-iron aqueduct
with a span of 65½ feet. The river Esla is a tributary of
the Douro; and the canal supplied by it can irrigate 32,000
acres, the water being directed into it by a masonry weir
across the river, 573 feet in length.

**Irrigation Canals in the United States.** Though primitive
methods of irrigation were employed from remote periods by

---

[1] 'Les Irrigations en Espagne,' André Llaurado, Annales des Ponts et Chaussées,
1878 (2), p. 617.

[2] 'Irrigation in Spain; the Henares and the Esla Canals,' G. Higgin, Minutes
of Proceedings Institution C. E., vol. xxvii. p. 489, and plate 22.

the Mexican inhabitants of Colorado and California, and about 1860 the American settlers began to divert water from neighbouring streams to their lands, irrigation in the dry belt of North America, by means of perennial canals, has only been undertaken on a large scale since 1872, when the first large irrigation canal was constructed, drawing its supply from the river Fresno in California, by the aid of a timber dam, 311 feet long, raising the water-level 6 feet [1]. The head of the canal, encased in a timbered channel 30 feet wide, is closed by six sliding gates. This canal, with a bottom width of 20 feet, side slopes of 2 to 1, and a depth of 8 feet, has a gradient of 1 in 10,560; and with a depth of water of 6 feet, and a velocity of 2 feet per second, was designed to discharge 360 cubic feet per second.

The Calloway Canal, commenced in 1875, takes its supply from the river Kern in California, which is fed by the snows of Mount Whitney; and the water of the river is kept up during the low stage by closing an open timber-work weir across the river, 400 feet long, with two-inch planks, which are removed on the approach of floods. The regulator of this canal is similarly constructed; and the canal, with a bottom width of 80 feet and depth of 5 feet, has been given side slopes of 3 to 1, owing to the light nature of the soil which it traverses. This canal, with a maximum flow of 700 cubic feet per second, running 3½ feet deep and with a velocity of 2½ feet per second, can irrigate 80,000 acres in the San Joaquin valley; and its gradient, which is 1 in 6600 for the first ten miles, is gradually reduced till it comes level at the end of its course of 32 miles, in order to reduce its discharge as the distributing canals branch off.

**Del Norte and Highline Canals, Colorado.** The Del Norte Canal, drawing a maximum supply of 2100 cubic feet per second from the Rio Grande at the town of Del Norte, for

---

[1] 'American Irrigation Engineering,' H. M. Wilson, Transactions of the American Society of Civil Engineers, 1891, vol. xxv, pp. 170 to 199.

irrigating about 225,000 acres in the San Luis valley, has a timber weir across the river at its head, similar to the Calloway Canal weir. This canal, which is 50 miles long, follows the slope of the ground for the first twelve miles, with an average gradient of 1 in 660, and without any falls on it ; but fortunately the coarse gravel and rock through which the canal was excavated, are firm enough to withstand the scour of the current without protection. When the canal leaves the hills, its slope is reduced to 1 in 2112. The regulator of this canal, constructed of timber-work, is closed by ten gates which, when raised, leave a total width of opening of 59½ feet with a height of 8 feet; and the canal has a bottom width of 60 feet, and a depth of water of 5½ feet.

The Highline Canal, diverging from the south channel of the river Platte 21 miles above Denver, is 85 miles long ; but being only designed to irrigate 90,000 acres, this canal is smaller, and has little more than half the discharging capacity of the Del Norte Canal, whilst its average gradient is 1 in 3000. The diversion weir across the river is formed of crib-work filled with rubble, constructed in 1884, in place of a less substantial weir which had been washed away ; and this weir is 117 feet long, and has a maximum height of 14 feet. This canal, just below its head, passes through a tunnel 600 feet long, and then along a timber channel or flume, 28 feet wide and 7 feet deep, skirting the steep rocky slope of the Platte cañon for half a mile ; and some miles further on, it is carried over a creek in another flume, 918 feet long, supported on high timber trestles.

**Arizona Canal, Arizona.** Irrigation by means of canals has also been extended to Arizona, by the construction of the Arizona Canal, which takes its supply from the Salt River 25 miles above Phœnix. The weir across the river at the head of this canal, 916 feet long and 10 to 11 feet high, is formed of crib-work filled with rubble ; a scouring sluice closed by planks has been placed between the weir and the

regulator, to keep the head of the canal free from silt; and
the influx into the canal is controlled by eight wooden lifting
gates, 4½ feet wide and 6 feet high. The main canal is
41 miles long; it has a bottom width of 36 feet; and with
a slope of 1 in 2640, and a depth of water of 7½ feet, it has
a discharge of 1000 cubic feet per second. With 125 miles
of branch canals, the Arizona Canal can irrigate 77,000 acres;
whilst two auxiliary canals at a lower level, 70 miles in
length, with 75 miles of branches, command 73,000 acres
of land.

**Bear River Canal, Utah.** The Bear River Canal, drawing
its supply from the Bear River 3½ miles above Colliston, with
a total length of principal canals of 150 miles, and capable
of irrigating 200,000 acres, is one of the most recent irrigation
canals constructed. The diversion weir, 370 feet long,
17½ feet high, and with a maximum bottom width of
38 feet, is formed of crib-work filled with earth and rubble;
and a canal branches off above it from each side of the river.
The regulators at the head of these canals have each five
openings, closed by iron gates, 4 feet wide and 7 feet high,
resting against masonry piers and abutments founded upon
the solid rock, so that these works are much more solidly
built than in the earlier canals. The canals pass along deep
cuttings in rock, through tunnels, and across deep ravines on
high embankments in the earlier portion of their course; and
the canal on the western side is carried across the Malad
valley in a wooden flume supported by an iron girder bridge
resting on iron trestles, 378 feet long, with a span of 70 feet
across the river Malad, and a maximum height of 80 feet.
The main canals of this system, with a slope of 1 in 5280,
a bottom width of 50 feet, and a depth of water of 7 feet,
discharge 1000 cubic feet per second.

**Idaho Canal, Idaho.** The Idaho Canal, drawing its supply
from the river Boise 12 miles above Boise City, is chiefly
remarkable for the solidity of its head-works, and the

extensive area which it can irrigate, amounting to 350,000 acres between the rivers Boise and Snake. The old channel of the river has been permanently closed by a solid dam of rubble, with earthwork on the up-stream side, 220 feet long and a maximum height of 43 feet, resting upon the rocky river-bed; and a side channel has been formed on the opposite bank to the canal, across which a masonry weir has been built, founded on a ledge of basalt, 500 feet long, and with its crest 8 feet below the top of the dam, over which the flood waters of the river pass. A sluice at the end of the dam adjoining the head of the canal, maintains the approach to the canal free from silt. The regulating gates, consisting of rolling-up curtains like those of the Seine weirs (p. 128), with steel laths below and wooden ones above, close openings 8 feet wide between masonry piers, and 19 feet high to the crown of the overhead semicircular arches.

Turlock Canal, California. The Turlock Canal, for irrigating 176,000 acres of the San Joaquin valley in central California, with a supply of water derived from the river Tuolumne, is specially remarkable for its diversion weir, which consists of a masonry dam, similar in section to the Vyrnwy dam in North Wales, 103 feet high and 310 feet long between the rocky walls of a cañon, situated 2 miles above La Grange[1]. The dam is founded on solid rock; and the maximum pressure on it, with the river rising to its crest, is 6.3 tons per square foot; and the river when in flood passes over its crest. As the river might rise in this narrow cañon to a maximum height of 16 feet above the sill of the canal, a tunnel, 500 feet long, with a gradient of 1 in 220, has been made through the hard rock for the entrance to the canal; and the regulating gates have been placed where the open canal commences at the lower end of the tunnel, to reduce the cost of these works. The six gates slide against angle-irons let

[1] 'American Irrigation Engineering,' H. M. Wilson, Transactions of the American Society of Civil Engineers, 1891, vol. xxv, p. 193.

into the rock, and close openings 3 feet wide and 12 feet high. The canal, excavated at first in slate rock, starts with a bottom width of 20 feet, a depth of water of 10 feet, and a gradient of 1 in 666, and discharges 1,500 feet per second with a velocity of current of $7\frac{1}{3}$ feet per second. The canal is then carried across ravines and valleys by means of dams and flumes, and through projecting spurs in tunnels; and for the first eighteen miles, the canal simply serves to convey the water to the lands to be irrigated, the section of the canal being eventually increased to a bottom width of 70 feet, side slopes of 2 to 1, a gradient of 1 in 5,280, and a depth of water of $7\frac{1}{2}$ feet. The main canal, after a course of 18 miles, divides into four principal branches with a total length of 80 miles, from which the water is distributed by 180 miles of minor branches, designed so as to have a uniform flow of $2\frac{1}{2}$ feet per second, in order to carry the suspended silt on to the land.

**Remarks on Upper Perennial Canals.** In the construction of these canals, as in other irrigation works, India occupies the foremost position; a condition imposed upon her by the density of the population, the dependence of most of the crops on irrigation, the vastness of the areas to be irrigated, and the abundant supply of water afforded by the large rivers which rise in the lofty snowclad ranges of the Himalayas. Irrigation by perennial canals was introduced into Spain at a very remote period; it was resorted to in Italy many centuries ago, and more than three centuries ago in the south of France; but the system has been extended in these countries in recent times, by the construction of the Henares and Esla canals in Spain, the Cavour Canal in Italy, and the Marseilles and Verdon canals in France. Almost all the large upper perennial canals of India belong to the nineteenth century; and even those which originated under native rulers, have been so transformed as to constitute new works. In America, perennial canals are quite recent works; and they have been developed to a remarkable extent within a quarter of a century.

The canals in India traverse comparatively flat country; and the chief difficulties in their construction have been the shifting nature of the rivers at their head, the long weirs required across the wide channels, and the alluvial soil in which the head-works have had to be founded; whilst the maintenance of the canals has been prejudiced by the tendency of the rivers to block up their entrances with sediment, and the difficulty of preventing the deposit of silt in the canals, owing to the reduced velocity of flow on entering the canal necessitated by the small fall generally available, and the danger of erosion of the bed in the alluvial soil by any increase in the gradient where feasible. The canals in northern Italy and the south of France draw their supplies from somewhat more stable rivers than the Indus and the Ganges; but the Marseilles and Verdon canals, and to some extent also the Cavour Canal, have their head-works at a long distance from the lands to be irrigated. The canals in the United States are under directly opposite conditions to those in India, with the exception of the first twenty miles of the Ganges Canal; for whereas their head-works, being in rocky districts, possess solid foundations, and are situated on rivers with a fixed channel, the canals have to traverse hilly country for the first few miles of their course before reaching the plains needing irrigation. These canals, however, being excavated in firm high ground, can be given an ample fall and a reduced section, and are more favourably situated than canals like those of Marseilles and Verdon, which, besides involving difficult head-works, have to traverse about fifty miles of most rugged country, before reaching the lands where their supply of water is utilized.

The largest of these perennial canals, in size of channel, discharge, and area irrigated, are the Ganges, Lower Ganges, Sirhind, and Sone canals; and next come the Bari Doab,

Western Jumna, and Cavour canals. The Chenab, Eastern Jumna, Idaho, Agra, Del Norte, Bear River, Pecos in New Mexico, Sidhnai and Sirsa in the Punjab, and Turlock canals command areas of 400 000 to 104,000 acres, with discharges of 2,500 to 1,000 cubic feet per second; whilst the other canals referred to above, irrigate areas of under 100,000 acres. The average gradient of these main canals, in open cutting or embankment, ranges between 1 in 2,112 on the Del Norte Canal, and 1 in 2,640 on the Idaho and Arizona canals, to 1 in 10,560, or 6 inches in a mile, on the Lower Ganges, Sone, and Agra canals.

Cost of Upper Perennial Canals. The cost of these canals, per acre that can be irrigated, necessarily varies considerably according to the conditions of the site of the head-works, the distance and nature of the country which the canal has to traverse before distributing its supply, the volume of water that can be conveyed, and the extent of land that can be commanded by the canal. As a general rule, canals conveying a large volume of water, and supplying an extensive area, cost less per acre irrigated than canals irrigating small areas ; and naturally canals with head-works, aqueducts, and bridges constructed largely of wood, as in America, are cheaper in the first instance than canals provided with the permanent works erected in India and Europe. The extremes of cost are a minimum of only 9s. per acre for the Idaho Canal, which must be remarkably favourably situated, and a maximum of £21 for the Verdon Canal, which, however, is also used as a source of water-power. These, however, are very exceptional cases; and the cost of the chief systems of irrigation canals described, ranges between 18s. and £8 7s. per acre irrigated, the cost of the Eastern Jumna and Sidhnai canals being at the lower limit [1], and the Cavour Canal reaching the higher limit of cost. The Bear River and Pecos canals cost £1 1s. per acre irrigable, the King's River and San

[1] 'Irrigation Works in India and Egypt,' R. B. Buckley, pp. 272 and 276.

Joaquin Canal £1 10s. [1], the Alpines Canal £1 15s., and the Ganges Canal £1 18s. The canals whose cost per acre irrigated is comprised between £2 and £3, are the Western Jumna Canal £2, the Calloway and Arizona canals £2 2s., the Sirsa Canal £2 8s., the Chenab and Lower Ganges canals £2 10s., the Highline Canal £2 14s., the Sone Canal £2 16s., and the Turlock and Bari Doab canals £3 per acre. The canals whose cost per acre irrigated lies about midway between the extreme limits, are the Esla Canal £3 2s., the Agra Canal £3 18s., the Alphonso Canal in Spain £4 5s., the Sirhind Canal £4 12s., and the Urgel and Tamarite canals in Spain £5 8s. [2]; whilst the canals approaching to the cost of the Cavour Canal are the Henares Canal costing £7 7s., and the Carpentras Canal £7 8s. per acre. In some instances, however, the cost has been increased by making them available for navigation; notably in the case of the Ganges Canal, as well as the Sone and Sirhind canals.

Most of the earlier irrigation canals proved unremunerative to their promoters, though of great service to the lands they irrigate; and the same has been the case with the Cavour and Verdon canals. The Indian perennial canals described have, however, yielded altogether a good average return, amounting in 1890-91 to a net revenue of between 5 and 6 per cent. on the capital expended. The receipts of the Sone and Chenab canals, indeed, were but slightly in excess of their working expenses; but the Eastern Jumna Canal gave a profit of 22 per cent., the Sidhnai Canal nearly 11 per cent., the Western Jumna Canal 9¼, the Bari Doab Canal 8, the Ganges Canal 7, the Sirhind and Agra canals 4, and the Lower Ganges Canal 2 per cent. Several also of the American irrigation canals should yield a good return, judging by their moderate capital cost per acre irrigated; but the expenses of maintenance of their works must prove considerably

---

[1] 'Manual of Irrigation Engineering,' H. M. Wilson, p. 70.
[2] Minutes of Proceedings Institution C.E., vol. xxvii, p. 520.

heavier than those of the solidly constructed works of the canals of India.

## DELTAIC PERENNIAL CANALS.

The delta of a river is a flat alluvial tract of land (p. 174), lying for the most part somewhat lower than the water-level of the river above the head of its delta, and of its diverging branches which intersect the delta, their beds and banks having been raised more rapidly by the deposit of a portion of the silt brought down by the river, than the land further off from these branches (Plate 5, Figs. 1, 2, and 5). Accordingly, the low-lying land between the diverging branches traversing the delta of a river, is readily irrigated by canals issuing from the river at the head of its delta, and passing along the higher ground near the margin of the branches, from whence the distributing canals branch off. Deltaic canals, therefore, passing through a flat district, and being situated close to the lands which they irrigate, present very little difficulty in construction compared to perennial canals traversing the higher parts of a river valley; and their head-works must be placed near the head of the delta, generally a little above the point where the branches separate, and where the river, having received the flow of all its tributaries, always discharges a fair volume of water in the dry season. These canals, moreover, having a very moderate fall, are not subject to erosion, and are suitable for navigation.

**Works for Deltaic Canals.** The head-works of these canals are similar to those already described in relation to perennial canals (pp. 434 to 436), consisting of a weir across the river to keep up its water-level, and regulating sluices at the head of the canal to control the influx. As a river after a long course, discharging the combined flow of all its affluents, is generally wide above the head of its delta, the weir is often of considerable length, as, for instance, the weir across the river Godavery at the head of the canals of the Godavery delta, which is nearly $2\frac{1}{2}$ miles long (Plate 3, Fig. 14); whilst the weirs, or regulating sluices, across the Rosetta and Damietta branches of the Nile delta, close to its head, are 1,437, and 1,709 feet long respectively, and are joined by a wall 3,280 feet in length, in the centre of which the head of the delta canal is situated.

The construction of the canals themselves is comparatively simple, for they pass through a flat and purely agricultural district, with a small uniform fall, and an alluvial soil unintersected by streams; whilst the course of these canals is fairly defined by the position of the branches of the delta. In rendering, however, this fertile land available for a succession of crops by irrigation, it becomes necessary to protect it from inundation, and also to drain it efficiently, in order to prevent the land becoming sodden with water. Embankments, accordingly, are carried along the branches of the river to exclude the floods from the irrigated lands; and the heads of these branches require protection, and the flow down them must be regulated, so that the changes which frequently occur at the head of a delta may not occasion an excessive influx into any of the branches, leading to the overtopping of the embankments and the inundation of the land. The protection, however, of the lands from inundation by embanking the branches, though enabling more crops to be raised, exclude also the fertilizing silt which used to be spread over the lands by floods; and this silt, being confined within the branches of the river, is liable to be partially deposited in their beds, raising their level more rapidly than before; and the remainder, being carried to their mouths, increases the rate of advance of the delta. The irrigated land, moreover, not merely loses the benefit to the crops of the periodical renewal of the layer of silt, but it ceases also to be gradually raised as the water-level in the branches rises with the prolongation of the delta, so that in time its protection from floods, and its efficient drainage, become increasingly difficult. The permanent success, accordingly, of deltaic irrigation depends upon the proportion of silt that can be carried on to the lands from the canals, without endangering the maintenance of their channels. Nevertheless, in spite of these prospective disadvantages, some extensive systems of deltaic irrigation have been constructed in India, three of which have proved very successful financially; whilst the delta of the Nile has been provided with irrigation canals, with somewhat unsatisfactory results.

**Deltaic Canal Systems in India.** Several rivers of India flowing into the Bay of Bengal and the Indian Ocean, form extensive deltas, namely, the Ganges, the Mahanuddee, the Godavery, the Kistna, and the Cauvery, which, with the exception of the Ganges delta, have been provided with irrigation canals.

The Orissa Canals, with 252 miles of main and branch canals, irrigate nearly 400,000 acres of the Mahanuddee delta, with a maximum discharge of 6,060 cubic feet per second, having been brought into operation in 1869. These canals, however, having cost £8 2s. per acre irrigated, though very valuable as a protection against famine, are worked at a loss [1].

The Godavery system of canals, commenced in 1849, having a maximum discharge of 8,500 cubic feet per second, irrigates 612,000 acres of the Godavery delta; these canals cost £2 2s. per acre irrigable; and they yielded a net revenue of 12⅛ per cent in 1890–91. The Kistna delta system, brought first into use in 1855, can irrigate 475,000 acres with a maximum discharge of 8,100 cubic feet per second; and having been constructed at a cost of £3 2s. per acre irrigable, it returned a profit of over 13 per cent. in 1890–91. The Cauvery delta system, the most extensive of these deltaic canals, irrigating 919,500 acres, is one of the oldest of the perennial canals of India; and its extension, commenced in

[1] 'Irrigation Works in India and Egypt,' R. B. Buckley, pp. 272 and 276.

1836, was, next to the Eastern and Western Jumna canals, the earliest important canal work undertaken by the British government. The revenue derived from this system of canals exhibits a remarkable percentage of profit on the capital expended in its improvement since 1836 ; but this makes no allowance for the old works carried out by the native rulers, which irrigated two-thirds of the present irrigable area.

**Deltaic Canals in Lower Egypt.** From time immemorial the extensive delta of the Nile below Cairo, constituting Lower Egypt, was mainly irrigated, like Upper Egypt, by the system of basins filled from inundation canals during the yearly flood of the Nile, so that every year the low lands were fertilized by the layer of silt deposited from the flood water. Cotton was only grown upon the higher lands which were only inundated at intervals of fifteen or twenty years, on the occurrence of an exceptional flood of the Nile. About the year 1820, however, Mehemet Ali introduced summer irrigation into the delta by excavating some of the old silted-up channels, and forming new canals, to a sufficient depth to receive the summer flow of the Nile, and thus enable water to be raised on to the land for the cultivation of cotton in the summer months [1]. This necessitated the raising of the embankments along the river and the canals, to protect these crops from inundation. The protected land, however, being deprived of its annual layer of rich mud, has greatly deteriorated ; whilst two crops are taken off it each year. Silt, moreover, accumulated rapidly in these deep canals, involving a great amount of forced labour each year to clean them out so as to admit the summer flow. This deposit also was promoted by regulators having had to be placed at intervals of 8 to 10 miles along the canals, owing to their large section and depth, in order that by being closed during the flood, they might raise the water to the surface of the

---

[1] 'Egyptian Irrigation, W. Willcocks,' pp. 49 and 146 ; and Minutes of Proceedings Institution C. E., vol. lxxxviii, p. 300.

land ; and in the absence in most cases of locks alongside these regulators, the navigation has been stopped. In addition, the great size of the canals facilitated so greatly the discharge of the floods, that the lands nearer the sea were inundated ; and the drainage of the cultivated lands having been impeded, the deposit of salt on the land increased.

In order to raise the level of the water during low Nile, and to regulate the supply, the barrages, or regulating sluices, were commenced in 1842 across the Rosetta and Damietta branches close to the head of the delta, which were only completed in 1861 ; but as on closing the segmental gates of the Rosetta barrage (p. 118) for the first time in 1863, a settlement occurred, these barrages remained open till 1890, when the completion of the repairs and extension of the foundations, commenced in 1887, enabled the numerous openings, $16\frac{1}{2}$ feet wide, between the piers of these barrages to be closed by vertical sluice-gates resting against rollers in the grooves. The water-level of the river at a low stage has thus at last been permanently raised, thereby enabling the yearly deepening of the canals to be dispensed with, and also storing up and regulating the summer discharge of the river. The drainage of the lands is being undertaken to reduce the salt deposits and to dry the swamps ; and a return periodically to basin irrigation for some of the lands in Lower Egypt has been advocated, to renew the soil by the deposit of the Nile mud ; whilst locks are being introduced to provide a passage for navigation where the canals have been blocked by the regulators.

**Remarks on Deltaic Canals.** These canals, irrigating special tracts of flat land in hot districts more or less deficient in rainfall, do not possess the same interest, or afford the same scope for skilful design and extension of irrigation, as upper perennial canals. Moreover, though the unsatisfactory results of the introduction of perennial canals for irrigating the delta of the Nile may be traced to the formation of too large and deep canals, unassisted by the raising of the low-water level of the river by a weir at the head-works; nevertheless, the exclusion of the flood waters, so necessary for the gradual raising of the level of the low-lying alluvial lands of a delta, as well as for renewing the soil exhausted by a succession of crops, renders this system of irrigation defective, as compared with irrigation by perennial canals in the higher portions of river basins. Possibly periodical inundation of the protected lands, as proposed for Lower Egypt, may prevent the exhaustion of the soil; but as this would only take place at intervals of some years, it cannot effect the same gradual raising of the low-lying land that a yearly deposit of silt produces. The process of raising the land by successive thin layers of silt is undoubtedly slow, depending upon the flooded area and the amount of silt carried on to the land, so that this defect in the system would be slow in manifesting itself to an appreciable extent. Nevertheless, in a river bringing down a large proportion of silt, and extending its delta rapidly, the cessation of the deposit of a yearly layer of silt on the protected lands, and the consequent more rapid raising of the bed and banks of the branches of the river through the delta, and the quicker advance of the delta, must in time prejudice the drainage of the protected lands, and render them more liable to a sudden inundation. Accordingly, though this system appears to have been carried out in the Cauvery delta for a very long period without injury, it cannot be regarded as a method of irrigation of assured permanent value in every case; though, provided the works are judiciously designed, its deficiencies, even under somewhat unfavourable conditions, may require the lapse of a long period of years to become manifest, owing to the slow rate at which a delta is raised by the yearly deposit of silt, and the possibility of bringing some of the silt upon the land with the irrigating water.

**Concluding Remarks on Irrigation.** Irrigation has been remarkably extended in India during the nineteenth century, and during recent years in the United States; whilst, even in the parts of southern Europe where it has been practised for some centuries, fresh canals have been constructed. In India, there is still scope for further important irrigation canals, especially within the basin of the Indus; whilst in America, there are very large tracts requiring irrigation, either by means of perennial canals, where sufficiently large rivers are available, or by numerous reservoirs in suitable localities. In Spain, some large rivers might be very advantageously utilized for irrigating fertile plains; whilst in Egypt, the birthplace of irrigation, the system of irrigating Lower Egypt is being transformed, and extensive works for the further development of irrigation in Upper Egypt are about to be commenced. In Africa and Australia, irrigation is still in its infancy; and large irrigation works will be required in the future to develop the agricultural resources of the extensive arid tracts in those countries.

Several irrigation works have proved unsuccessful financially; but with increased scientific knowledge and enlarged experience in the execution of public works, together with exhaustive surveys, observations of rainfall, and gaugings of the discharge of streams and rivers, such as are being carried out by the geological survey department of the United States, past errors should be avoided, and correct estimates formed of the prospects of schemes for irrigation. Moreover, as irrigation furnishes the only means of extending cultivation to arid regions, and also greatly increases the productiveness of the soil in hot dry climates, it will gradually assume a much greater importance as the population of the world increases, and the available fertile lands have become occupied, for the population depends upon the food supply that can be raised by the land.

# CHAPTER XX.

## INLAND NAVIGATION.

Importance of Inland Navigation in Continental Europe. Influence of Railways on Inland Waterways. Importance of Waterways penetrating far Inland; instances of large rivers. Conditions affecting Construction of Canals. Traffic on Inland Waterways dependent on Trade in Bulky Goods, examples in France and England. Canals connecting River Navigations, examples in flat country, also in hilly districts; Importance of favourable Conditions. Inland Navigation in relation to the Countries of Europe. *Inland Navigation in Great Britain :*—Development; Causes of Decline; Waterways with large Traffic, causes of success; Hindrances to Through Traffic; Aids to Inland Navigation, expedient methods; Lengths of Waterways in England, Scotland, and Ireland; Traffic Returns, incompleteness of, for England, Scotland, and Ireland. *Inland Navigation in France :*—Early Progress; Briare and Languedoc Canals; Extension of Canals; Progress since 1860, works, main routes rendered uniform; Lengths of French Waterways; Traffic, compared with Railway Traffic. *Inland Navigation in Holland :*—Natural Facilities; Principal Waterways; Lengths of Dutch Waterways; Control; Sizes, new Merwede Canal; Remarks. *Inland Navigation in Belgium :*—Natural Advantages; Principal Waterways, changes of level; Lengths of Belgian Waterways, Sizes; Traffic, nature and improvement; Traffic on Belgian Railways and Waterways compared. *Inland Navigation in Spain :*—Navigable Rivers; Castile and Aragon Canals. *Inland Navigation in Italy :*—Development, and Principal Waterways; Lengths and Sizes of Italian Waterways, Locks; Control; Traffic.

THE development of inland navigation is regarded as a matter of national importance on the continent of Europe; and it is undertaken by the State in North America and India. In France more especially, the completion of the connections between the principal waterways of the country is being steadily carried out by the Government; and the main lines of inland waterways have been given uniform dimensions, to enable vessels of 300 tons to pass along them throughout their whole length. Waterways are, indeed, regarded in France as essential routes for traffic, which must be made and maintained out of the taxes for the public benefit, and be free of tolls, similarly to roads; and similar views as to inland waterways, are held by the other continental countries of Europe, though the connection between the various waterways has not been developed as much in other countries as in France.

**Influence of Railways on Inland Waterways.** The introduction of railways, by providing a novel competitor with waterways, and rapid and regular means of transit, naturally checked for a time the further development of inland navigation; but after a few years, it was generally acknowledged that waterways are very useful auxiliaries for certain classes of traffic, and that by offering an alternative method for the conveyance of bulky goods, they are very valuable in keeping down the rates on railways. In England and the United States, however, the canals, in some cases, were bought up by the railway companies; and no efforts were made, in many instances, to increase the accommodation afforded by the canals, and to secure uniformity of gauge on the several links of connected waterways, so as to meet the growing demands of trade, and compete with railways for the traffic. Railways gave a great stimulus to trade, and increased the facilities of transport; and those inland waterways have proved the most prosperous, which passing through mineral districts or places possessing a trade in bulky goods, and communicating with populous centres or seaports, have been gradually improved and enlarged to conform with the requirements of commerce.

**Importance of Waterways penetrating far Inland.** The value of inland navigation, under ordinary conditions, depends to a great extent on the distance to which it can penetrate into the interior of a country, and thus by a long unimpeded course, compete on the most favourable terms with railways for traffic between the interior and the sea-coast. Water-carriage, moreover, is most economically conducted by vessels of large capacity in an unrestricted waterway. Accordingly, large rivers possessing a good navigable channel for a long distance inland, such as the Rhine, the Danube, the

Mississippi, and the Amazon, furnish valuable natural highways for inland navigation. These rivers, moreover, and others less favourably circumstanced by nature, can be improved by regulation works, like the Elbe and the Oder, and their navigable capabilities extended further inland. Many rivers, also, with a less constant discharge, or with a less extensive drainage area, have been rendered navigable throughout the year by canalization a long way into the interior, as described in Chapter III. The value of such waterways for traffic is exemplified by the flourishing ports of Mainz and Mannheim [1] on the Rhine, in the interior of Germany, 306 and 352 miles respectively above Rotterdam, and also by the great increase in navigation on the Lower Seine between Rouen and Paris, and on the Main between its confluence with the Rhine, opposite Mainz, and Frankfort, by the canalization of these rivers, notwithstanding, in the latter case, the competition of a railway along each bank of the river.

**Conditions affecting the Construction of Canals.** Canals are introduced to supply the deficiencies of natural waterways, by being constructed in places where rivers are not available, or to provide a passage where serious obstacles to navigation exist in a river, or to connect two river navigations by surmounting the water-parting of their basins. As canals can be cheaply constructed in flat countries, such as Holland, Flanders, and Lombardy, extensive use is made of them in such districts. Under less favourable conditions, the construction of canals where rivers are not suitable for navigation, must depend upon the nature and extent of the traffic of the district they are intended to serve, and whether they can be made to connect important inland towns with seaports at a reasonable cost.

The employment of lateral canals, in place of portions of a river, is generally due to economical considerations, as being

---

[1] 'Canal, River, and other Works, in France, Belgium, and Germany,' L. F. Vernon-Harcourt, Minutes of Proceedings Institution C.E., vol. xcvi, pp. 198–200, and plate 6, figs. 7, 8, and 10.

cheaper in certain cases than the improvement of the river itself; and in some instances the adoption of a canal is essential, as in the case of the Welland Canal in place of the Niagara River, since it forms a link in the most important inland navigation system of the world. Where a canal wholly replaces a river, like the Loire lateral canal, its execution is determined by the same conditions as an independent canal, bearing in mind the advantages it possesses of running along the valley of a river, and deriving its supply of water from the river.

The expediency of constructing canals to connect two river navigations depends upon the prospect of through traffic between the places served by the existing navigations, and the cost of a canal traversing the interval separating the two navigations, which is frequently long, and having to surmount the ridge separating the basins. Connections of this kind have the advantage of linking the waterways together; but sometimes they are not commercially profitable, owing to the large outlay involved in relation to the traffic frequenting such a route. Uniformity in the available dimensions of the waterways forming these through routes, is essential to the due development of the traffic along them; but this consideration was frequently overlooked in the construction of the various links by different companies, so that sometimes one link of inadequate dimensions hampers the navigation over a long length of waterway.

**Traffic on Inland Waterways dependent on Trade in Bulky Goods.** Large rivers affording an ample and unrestricted waterway for long distances inland, and canals which have been enlarged to meet the requirements of trade, exercise an important influence on the prosperity of inland navigation; but much the most powerful factor in the success of inland navigation is the nature of the traffic, and the position of industrial centres in relation to seaports. A large trade in coals, minerals, timber, or other bulky goods, between a district in the interior and a seaport or large city, is most favourable to the development of a large traffic along a suitable waterway; and in fact, in most cases, the bulk of the traffic is mainly confined to those waterways which possess this sort of trade. Thus in France, the waterways with a large traffic are the Upper and Lower Seine, bringing materials to Paris; and more especially the rivers and canals connecting Paris, Dunkirk, the coalfields of the northern corner of France, and the adjacent coalfields and mineral districts of Belgium [1]. In England also, the traffic returns are the largest on the Birmingham Canals, the Aire and Calder Navigation, the Bridgewater Canals, the Leeds and Liverpool Canal, and the Weaver Navigation, all of which, owing to their position, possess a large and profitable trade in bulky goods. More-over, though the successive enlargements of the Aire and Calder Navigation, and the attention paid to traffic arrange-ments along it, have rendered it the most prosperous waterway in England, and the increased facilities afforded along the Weaver Navigation, only 20 miles in length, for its large traffic in salt, have maintained it in a very satisfactory position, smaller improvements on the Bridgewater Canals have enabled these waterways to secure ample returns on account of the favourable nature of their traffic. The Birmingham Canals also, in spite of their antiquated inadequate dimensions, possess a revenue not much inferior per mile to that of the Weaver Navigation, owing to their exceptional position. In fact, whilst the timely enlargement of a waterway, and increased facilities for traffic greatly promote the extension of inland navigation under favourable conditions, it is difficult, even by neglect of requisite improvements, to drive away a traffic in bulky goods from waterways suitably situated in the midst of industrial centres in the interior; and, on the other hand, it is impossible to procure a large traffic along convenient waterways in districts where no trade in bulky goods exists.

[1] 'Statistique de la Navigation intérieure.' Ministère des Travaux Publics, Paris, 1894, vol. ii, plate.

**Canals connecting River Navigations.** Before the introduction of railways, it was necessary to connect river navigations by a canal surmounting the water-parting of their basins,
in order to obtain through routes between important districts
situated in different watersheds. Several of these connecting
canals were, accordingly, carried out during the development
of inland navigation (p. 348); some, moreover, have been
constructed within recent times, particularly in France, where
the completion of the communications between the main
waterways is considered essential; whilst almost the only
canals in Russia have been formed with the object of connecting the large rivers of that country. The value of such
connections under favourable conditions, in the case of rivers
navigable for long distances, is unquestionable; as for instance
in Russia, where the comparatively low ridges separating the
basins are easily surmounted, and where a fairly short canal
unites by water distant parts of that vast empire. A still
more remarkable example is the Obi-Yenisei Canal in
Siberia, which passing through a marshy region for a length of
about 5 miles, with a depth of cutting not exceeding $17\frac{1}{2}$ feet,
connects two tributaries of these great rivers, and thus has
opened up a waterway with a continuous length of about 3,300
miles, extending from Tumene to Irkoutsk.

**Canals in Hilly Districts for joining Rivers.** In Western
Europe the circumstances are much less favourable, for the
river basins are smaller in area and less flat; and the rivers
therefore are smaller, shorter, and more rapid, whilst the
ridges separating their basins are at a higher elevation. Under
these conditions, the canal connecting the navigable portions
of two rivers across their water-parting is generally long, and
has to undergo considerable variations of level; whilst an
adequate supply of water is obtained with difficulty at the
high summit-level. Thus the Rhine-Marne Canal ascends
from the Rhine at Strassburg, by fifty-one locks, to a summit-
level 432 feet above the river, and then descends about 230 feet

into the upper valley of the Moselle, which it has to traverse in order to reach the ridge of the Marne basin ; and it crosses this ridge at a summit-level of 480 feet above its starting-point on the Rhine, previously to its final descent of 590 feet, by seventy locks, to form a junction with the Upper Marne Canal at Vitry-le-François, after a course of 191 miles. The Rhine-Rhone Canal, diverging from the Rhine above Strassburg, and running parallel to the Rhine southwards for about sixty miles to Mulhouse, where it turns towards the west to reach the Saône basin, rises 697 feet by eighty-five locks to its summit-level ; whence it descends by seventy-seven locks to join the Saône at St. Symphorien, after a course of 199 miles, and thus obtains access to the Rhone at Lyons where the Saône flows in. The Rhine-Marne Canal has a very good traffic both into and out of France, ranking in average amount of traffic in France next after the northern waterways and the Seine ; but the Rhine-Rhone Canal has a very small traffic, as there are better and more direct waterways between the Rhone and the north, so that in this case the great cost involved in forming a through route has not by any means been compensated for by the traffic obtained. The inadequacy of the traffic along this route might be attributed partially to its joining rivers in different countries, leading to absence of encouragement to trade ; but the Bourgogne Canal also, connecting the Rhone basin with the Seine basin and Paris, by passing from St. Jean-de-Losne on the Saône to Laroche on the Yonne, and situated wholly in France, possesses only a very moderate traffic, quite incommensurate with the cost of its construction. The Bourgogne Canal, indeed, has a total length of 150 miles, and rises 654 feet from the Saône, by seventy-one locks, to cross the ridge between the Saône basin and the Yonne basin, and then descends 980 feet by one hundred and thirteen locks to the Yonne. The Main-Danube Canal, forming a connection between two of the principal rivers of Central Europe, the Rhine and the Danube, should be a very important waterway; but it is hampered by the high elevation, and consequent considerable fall of the Main and the Danube where it joins them, resulting in a very inadequate depth in the dry season in these rivers, and an average depth of only 5¼ feet. Moreover, the canal, though starting from the Main at Bamberg at an elevation of 504 feet above the Rhine at the confluence of the Main, has to ascend 601 feet to its summit-level by sixty-eight locks, before descending 260 feet to the Danube by thirty-two locks, which it joins at a height of 1,112 feet above sea-level.

**Favourable Conditions required for Canals connecting Rivers.** In view of the competition of railways, with their superior facility for surmounting high ridges, it is evident that the construction of a canal to connect two rivers can only be commercially profitable when the rivers possess a good navigation, when the water-parting is at only a moderate elevation, where there is a need of through communication, and where the class of traffic is specially suitable for carriage by water. The conditions for the construction of a canal are always more or less unfavourable in the upper portions of river valleys, and the traffic is impeded by the number of locks; and therefore this class of canal can only prove advantageous when the conditions are specially favourable. Thus, in Great Britain, the Thames and Severn Canal, uniting two river navigations with a small traffic, and which have been left in a very imperfect condition, has a total rise and fall, in conjunction with the Stroudwater Canal, of 478 feet, though piercing the top of the ridge in a long tunnel, which is surmounted by fifty-seven locks in a length of 38 miles; and this canal, which originally cost £249,330, has very little traffic, and is worked at a loss. On the other hand, the Forth and Clyde Canal, connecting two deep estuaries adjacent to the two principal cities of Scotland, with a total rise and fall of 312 feet, surmounted by thirty-nine locks in a length of 39 miles, possesses a large traffic and a good revenue.

**Inland Navigation in relation to the Countries of Europe.**

Inland navigation is of most value in countries which have
a small sea-board in proportion to their area, and for districts
at a distance from the coast; and it is most readily developed
where extensive plains stretch far into the interior, as in some
parts of Germany and Russia. Large continents, also, possess
generally large river basins, so that the distance from the
sea-coast is somewhat compensated for by the large rivers
traversing extensive inland tracts, which provide natural
waterways, and on which the chief centres of population are
usually situated, such as the Danube, the Rhine, the Elbe,
the Oder, and the Volga and other large rivers of Russia.
In Europe, accordingly, Northern Germany, large tracts of
Russia, Hungary, parts of France, and the low-lying lands
of Holland and the north-west of Belgium, are specially suited
for inland navigation; whereas Norway, most of Sweden,
Turkey, the greater part of Spain, Austria except the valley
of the Danube, and Italy except the plains of Lombardy,
are not adapted by nature for inland navigation. Great
Britain also, with its great extent of sea-coast, its numerous
sheltered seaports, its limited area, and its small rivers, is not
favourably circumstanced for the development of inland waterways, except under special conditions.

### INLAND NAVIGATION IN GREAT BRITAIN.

The earliest canals in England, the Caer Dyke and the
Foss Dyke, were constructed by the Romans in the flat fen
districts, and improved in the twelfth century. Traces only
exist of the Caer Dyke, which was 40 miles long, and connected Peterborough with the Witham; but the Foss Dyke,
10½ miles long, joining the Trent at Torksey with the Witham
at Lincoln, is still navigable. The navigation of the rivers was
gradually improved, and canalization was slowly carried out
during the seventeenth century, towards the close of which the
Aire and Calder Navigation was opened. The Bridgewater
Canal, however, from Worsley to Manchester, 10¼ miles long,

carried out by Brindley under the instructions and with the capital of the Duke of Bridgewater, and completed in 1761, gave the first impulse to the development of the canal system in England, which was rapidly extended during the latter part of the eighteenth century and the early years of the nineteenth century, Brindley himself laying out 523 miles of canals.

**Causes of the Decline of Canal Traffic in England.** The last inland canal was completed about 1834, attention being thenceforward directed to the extension of railways, which resulted in a rapid decline in the prosperity of canals. This decline in the canal traffic was due in some cases to the superiority of railways in surmounting physical obstacles, such as were met with on the Thames and Severn route; and in the case of through routes, to the absence of uniformity in dimensions, the division of ownership and management, and occasionally the purchase of one of the links by a railway company. In most instances no effort was made to strengthen the canal system against the new competitor, by timely enlargements of the waterways, improved accommodation, and the provision of uniformity in gauge; the canal proprietors were often eager to sell their undertaking to the competing railway company, who thereby secured complete control of the district; and occasionally a canal was actually converted into a railway.

**Instances of British Waterways with large Traffic.** In comparatively few instances only, inland navigations have flourished in England in spite of the development of the railway system, owing to the peculiarly favourable nature of their traffic, aided in rare cases by successive enlargements and improved management. The large traffic and good revenue afforded by the Birmingham, Grand Junction, Leeds and Liverpool, Rochdale, and Bridgewater canals, and the Trent and Mersey and the Calder and Hebble navigations, are due to their position, and the bulky nature of their traffic in the Birmingham, Manchester, Staffordshire, and West Yorkshire districts, and with various collieries, for the available draught along them is only from 3⅓ to 4 feet; and the Birmingham Canals and the Trent and Mersey Navigation cannot admit barges exceeding 7 feet in width [1]. The Aire and Calder and the Weaver navigations, on the contrary, furnish notable examples of waterways which, possessing a large traffic in coal and salt, have been gradually enlarged, and have maintained the competition with railways with very satisfactory financial results. Thus, the locks on the Aire and Calder Navigation, which were originally 56 feet by 14 feet by 3½ feet, have now minimum dimensions of 215 feet by 22 feet by 9 feet; whilst the locks on the Weaver Navigation have been successively enlarged from narrow locks with 4½ feet depth in the eighteenth century, to 88 feet by 18 feet by 7½ feet in 1831, then to 100 feet by 22 feet by 10 feet in 1860, and lastly, only about seventeen years later, to 220 feet by 42 feet by 15 feet, enabling vessels of 1 000 tons to use the navigation in place of 40 tons originally [2]. The Don and the Forth and Clyde navigations also, are instances of waterways having a good depth and possessing a large traffic and a good revenue, for their available depths are 6¾ feet, and 9½ feet, though their locks, with minimum widths of 17 feet and 19 feet 10 inches, have the inadequate lengths of 61¼, and 68½ feet respectively.

**Hindrances to Through Traffic on English Waterways.** The most serious impediments to the progress of inland navigation along the main routes in England, are the differences in the dimensions of the waterways, and the variety of ownership. Thus Birmingham, the most important town in the centre of England, has four possible routes to the sea, namely by waterways to the Thames, the Humber, the Mersey, and the Severn; and as the surrounding district possesses a large trade in bulky goods, an ample waterway between

---

[1] 'Returns made to the Board of Trade in respect of the Canals and Navigations in the United Kingdom for the year 1888.'

[2] Minutes of Proceedings Institution C.E., vol. lxiii, p. 262, and plate 11.

Birmingham and the sea should secure a large traffic. The route, however, from Birmingham to the Thames at Brentford is owned by four independent companies; the largest of the waterways of this route, the Grand Junction Canal, can only pass vessels of 50 tons; whilst two of the canals traversed can only admit barges 7 feet wide and $3\frac{3}{4}$ feet draught, carrying less than 30 tons [1]. A scheme for the amalgamation of these canal companies has been recently started, which it may be hoped will result, not merely in placing the entire waterway between Birmingham and London under a single management, but also in a much-needed improvement of its navigable capabilities. The route from Birmingham to the Humber is also owned by four companies, two of them under railway control; and as the available draught along part of this route is reduced to $2\frac{1}{2}$ feet, it can only be traversed by barges carrying less than 20 tons. The waterway from Birmingham to the Mersey is owned by three companies, and is almost wholly under railway control; whilst it can only be navigated by barges carrying about 20 tons. The shortest means of access for Birmingham by water to the sea, and the one most capable of adequate improvement without much difficulty and at a moderate cost, is by the Worcester and Birmingham Canal to Worcester, and thence by the Severn and the Gloucester and Berkeley Canal to the Bristol Channel, as vessels of 400 tons can already get up as far as Worcester, only 30 miles from Birmingham. The canal, however, at present can only admit barges of 30 tons. Most of the other through routes by water in England are similarly hampered.

**Aids to Inland Navigation in Great Britain.** The purchase and improvement of all the waterways of Great Britain by the Government have been suggested as a remedy for the depressed state of its inland navigation; but considering the

---

[1] 'Inland Navigation, with special reference to the Birmingham District.' L. F. Vernon-Harcourt, Transactions of the Federated Institution of Mining Engineers, 1894-5, vol. viii, and plate 26.

extensive sea-board and the good tidal rivers of the country, the convenient accommodation provided by railways, and the limited extent of the inland districts to be served, the vast expenditure which this course would entail could not be justified. Moreover, whilst some waterways would amply repay a judicious outlay on improvements, others could never be expected to yield any profit; and yet all riparian owners would consider that they had equal claims upon the assistance of the Government. The Government, however, might with great advantage facilitate the amalgamation of companies owning the several links of important through routes, and the release of canals from the control of railway companies; and they might also usefully grant public loans at a low rate of interest for the consolidation and improvement of waterways of acknowledged value. The recent purchase of the Don Navigation by a Sheffield syndicate from the Manchester, Sheffield, and Lincolnshire Railway Company, and the proposed amalgamation of the companies owning the canals between Birmingham and London, are important steps in this direction; and a few more similar enterprises, and amalgamations of ownerships of through routes, would greatly promote the extension of inland navigation in England.

**Lengths of British Inland Waterways.** The combined length of inland canals and navigations in England is given in the Board of Trade return as 3,050 miles; but a later estimate increases this length to 3,511 miles, exclusive of ship-canals, 2,232 miles of which belong to independent companies, and 1,279 miles are under the control of railway companies [1]; whilst 135 miles are stated to have been converted into railways, and 210 miles of navigations abandoned. Scotland, with its numerous firths, its mountainous districts, and its comparatively small inland population, is not generally suitable for inland navigation; and according to the return, it

---

[1] 'Map of Canals and Navigable Rivers of England and Wales.' Lionel B. Wells, 1894.

possesses only 52½ miles of ordinary inland canals, all belonging to railway companies; but by including the Caledonian and Crinan canals, the total is raised to 122 miles. Inland navigation has been more developed in Ireland, being aided by the Shannon and other rivers; so that the length of canals and navigations in Ireland is 609½ miles, of which only the Royal Canal, 96 miles long, is owned by a railway company.

**Traffic on British Inland Waterways.** The Board of Trade returns of traffic along the inland waterways of the United Kingdom in 1888, are stated to be incomplete; and they furnish no record of the distances along which the traffic is conveyed, so that the number of ton-miles cannot be ascertained. Moreover, the traffic passing over two or more different companies appears in the returns of each of the companies, and is consequently repeated twice or oftener. The returns, therefore, of the tonnage do not afford an accurate indication of the relative traffic on the several waterways, nor do they enable a comparison to be instituted between the traffic on British and on continental waterways. The total tonnage, however, given in the returns amounts to 36,457,100 tons for the United Kingdom, if the traffic on the Caledonian Canal is included, of which 34,325,200 tons belong to England, 1,612,300 tons to Scotland, and 519,600 tons to Ireland, showing that the traffic along the waterways of Ireland is much less in proportion to their mileage than in England or Scotland, owing doubtless to the absence of mineral resources in Ireland, and the comparatively small extent of its manufactures.

## INLAND NAVIGATION IN FRANCE.

The rivers Ourcq, Vilaine, and Lot were canalized in the sixteenth century, by means of locks, which are supposed to have been first introduced at Spaarndam in Holland in the thirteenth or fourteenth century, and in Italy, near Milan, in the fifteenth century; and about 100 miles of canals had been formed in France by the close of the sixteenth century.

**Briare and Languedoc Canals.** The first canal of importance constructed in France, with locks and a summit-level across the boundary of two river basins, was the Briare Canal, 36¾ miles long, opened in 1642, which connected the Loire with the Loing, a tributary of the Seine, and thus united by water two of the largest rivers of France. The next large canal carried out was the Languedoc Canal, now called the Canal du Midi, which was authorized in 1666, and completed in 1681, and connects the Garonne at Toulouse with Narbonne near the Gulf of Lyons. This canal, 140 miles long, was formed with the ambitious object of enabling vessels to avoid the long detour by the Straits of Gibraltar, by providing a waterway across the south of France between the Bay of Biscay and the Mediterranean. The canal, however, though a remarkable work considering the period at which it was constructed, could only accommodate a very small class of vessels ; and, moreover, it had to rise from the Garonne by eight single, and nine double locks, to a summit-level at Naurouse, 610 feet above the level of the Mediterranean, to which it descended by seventy-two locks, with falls of 5 to 12 feet [1].

**Extension of Canals in France.** Canals were gradually extended in France ; and by the close of the seventeenth century, 415 miles of canals were in existence, which were increased to about 630 miles by the end of the eighteenth century [2].

The opening of the Canal du Centre in 1793, by joining the Saône and the Loire, completed the connection by water between the English Channel and the Mediterranean ; and the St. Quentin Canal, opened in 1810, provided direct access from this waterway to the North Sea. The main rivers were gradually connected during the earlier portion of the nine-

---

[1] 'Des Canaux de Navigation et spécialement du Canal de Languedoc.' J. de Lalande, 1778, p. 35.

[2] 'Verhandlungen der Allgemeinen und Abtheilungs-Sitzungen, III. Internationaler Binnenschifffahrts-Congress zu Frankfurt am Main,' 1888, p, 21.

teenth century, by the Bourgogne Canal, joining the Yonne and the Saône, in 1832, the Rhone-Rhine Canal in 1834, the Sambre-Oise Canal in 1839, the Nivernais Canal from the Yonne to the Loire in 1842, and the Marne-Rhine Canal in 1853. The construction also of lateral canals along the upper portions of some of the rivers which were not suitable for canalization, and the canalizing of the main rivers with the exception of the Rhone and the Loire, improved the waterways which, by means of the connecting canals, extended over long distances, especially in the central, eastern, and northern parts of the country.

**Progress of Inland Navigation in France since 1860.** On the introduction of railways, the waterways were for a time neglected in France ; and, as in Great Britain and the United States, some of the canals were conceded to the railway companies. A reaction in favour of water communications, however, took place in 1860, as providing a means of keeping down the railway rates, and also in view of the increased traffic that might be expected to result from the commercial treaties which had been entered into by France. The chief works undertaken were the improvement of the rivers, the canalization for instance of the Upper Seine having been commenced in 1860, and the further improvement of the Lower Seine having been undertaken in 1866; but the completion of some canals was also carried out at this period, and several of the conceded canals were repurchased, such as the Rhone-Rhine, Bourgogne, Bretagne, Nivernais, Loire lateral, and Berry canals[1]. The tolls, also, on navigation were reduced in 1860, and again in 1867.

The improvement works were necessarily arrested by the Franco-German war ; and the navigability of the through routes was considerably restricted by the differences in depth, and in the sizes of the locks, of waterways constructed at

---

[1] 'La Navigation intérieure en France.' D. Bellet, Association Française pour l'Avancement des Sciences, Paris, 1889, part ii, p. 1035.

different periods. The increased interest, however, in inland
navigation aroused about 1876, directed attention to the
impediments caused by this want of uniformity, and led to
the law of 1879 for making all the main lines of waterways
conform to specified standard dimensions (p. 376), and also
resulted in the construction of important connecting links.
In 1878, only 619 miles of rivers and 287 miles of canals were
of the standard dimensions, or altogether 906 miles ; whereas
by 1893, the works of reconstruction had increased these
lengths to 1,202 miles of rivers and 1,353 miles of canals,
giving a total of 2,555 miles of waterways in France possessing
the requisite dimensions [1]. Of the 1,066 miles of canals con-
forming to the standard dimensions added since 1878, 362
miles consist of new canals constructed for the most part,
either for the re-establishment in French territory of the
connection of waterways severed by the cession of Alsace
and Lorraine, effected by the Canal de l'Est, 224 miles of
which were opened since 1878, or to complete the links for
connecting waterways, of which the most important are the
extension of the Upper Marne Canal for 23⅝ miles, which
with the Marne-Saône Canal, 95 miles long, in course of
construction, will connect these two rivers, and the Oise-Aisne
Canal, 30 miles long, opened in 1889.

**Length of French Inland Waterways.** The total length of
waterways in France used for navigation in 1893, amounted
to 7,657 miles ; but as this length includes 528 miles of rivers
which are merely floatable, the actual length of navigable
rivers and canals is 7,129 miles ; namely 4,143 miles of rivers
and 2,986 miles of canals. The lengths of the canalized
rivers and of the open navigable rivers are about equal ;
whilst about four-sevenths in length of the canals connect
rivers across the water-parting of their basins, and the

---

[1] 'Statistique de la Navigation intérieure,' Paris, 1894, vol. i, p. 17 ; and
'Navigation Works executed in France from 1876 to 1891.' F. Guillain, Transactions
of the American Society of Civil Engineers, vol. xxix, p. 12.

remainder consist mostly of lateral canals. The whole of the navigable rivers of France are under the control of the State, except 6¼ miles of the canalized Lez; and only 491 miles of canals remain in the hands of private companies, of which the most important are the Garonne lateral Canal, 132 miles long, ceded till 1960, and the Canal dú Midi, 173 miles long, ceded in perpetuity to the Southern Railway Company, the Paris canals of Ourcq, St. Denis, and St. Martin, with a total length of 74½ miles, worked by the city authorities, and the Sambre-Oise Canal, 41⅜ miles long.

**Traffic on French Inland Waterways.** The total tonnage carried over French waterways in 1893 amounted to 25,101,500 tons, of which 10,523,900 tons were conveyed on the rivers, and 14,577,600 tons on the canals; whilst the number of tons carried one mile reached 2,203,778,000 tons, and the average traffic over the whole length of waterways, or the ton-miles divided by the length, was 287,812 tons[1]. The inland navigation traffic, however, is very unevenly distributed in France, for there is hardly any traffic along the inland waterways in the west and south of France, to the west of a line drawn from the mouth of the Seine to Cette on the Mediterranean; whilst the bulk of the traffic is concentrated on the Seine between Montereau and Rouen, the Marne-Rhine Canal, and the northern river navigations and canals which connect these waterways with Dunkirk and Belgium[2].

The Paris section of the Seine heads the list as regards absolute tonnage, with 4,758,500 tons, followed closely by the St. Quentin Canal and the Upper Deûle; whilst the Oise lateral Canal, the sections of the Seine just above and below the Paris section, and the 7 miles of the Scheldt between Cambrai and Étrun, possess a traffic with an absolute tonnage exceeding three million tons annually. Taking, however, the

---

[1] 'Statistique de la Navigation intérieure.' Paris, 1894, vol. i, p. 65, and vol. ii, pp. 129 and 189.

[2] Ibid. vol. ii, plate.

average traffic throughout each section, which furnishes a more exact basis for comparison, the largest amount passes along the section of the Scheldt between Cambrai and Étrun, reaching 3,485,200 tons in 1893, closely followed by 3,367,100 tons on the Lower Seine between La Briche and Conflans, and by 3,337,400 tons on the St. Quentin Canal; whilst the average traffic along the Seine between Montereau and Rouen, and along the Oise and the canals connecting it with Dunkirk, exceeds for the most part two million tons, only falling towards one million tons near the extremities, at some distance from Paris, towards which the traffic from the various waterways converges. The traffic on the Marne-Rhine Canal averages about one million tons; but to the south of this canal and of the Seine at Montereau, the traffic along the most frequented route, from the Saône to the Seine, by the Canal du Centre, the Loire lateral Canal, and the Briare and Loing canals, merely ranges between 443,800 tons and 662,600 tons; whilst the maximum traffic on the Saône between St. Jean-de-Losne and Lyons is only 336,400 tons, and on the Rhone between Lyons and Arles 223,300 tons. The traffic on the Garonne lateral Canal and the Canal du Midi appears to have been practically suppressed by the Southern Railway Company, which follows approximately the same route, and to which these canals belong; for these canals, which seem to provide an important through route between Bordeaux and Cette, and therefore between the eastern and southern coasts of France, have an average traffic from Castets nearly to Cette, along a distance of 270 miles, of only from 77,200 to 53,700 tons.

The railway traffic of France is much more evenly distributed than the traffic by water; and though it exhibits a similar concentration round Paris and to the north-eastern frontier and Havre, it is immeasurably larger than the traffic by water in the central, western, and southern parts of the country. Moreover, the superiority of the railway traffic is very marked south of Montereau to the Mediterranean; but this may be partly accounted for by the absence of connection by an inland waterway between the Rhone and Marseilles. The traffic by water, however, has increased since 1880 from 1,227,361,000 ton-miles to 2,203,778,000 ton-miles in 1893, owing no doubt at first to the abolition of tolls on the waterways in 1880, and subsequently to the improvements effected along the main routes.

## INLAND NAVIGATION IN HOLLAND.

Holland is specially adapted by nature for inland navigation, not merely on account of the flat low-lying character of its land, which renders the construction of canals very easy, but also owing to the number of large branches of the Rhine and the Maas, which intersect the southern part of the country, and the Zuider Zee, together with several inland lakes, which afford natural facilities for internal transport by water. The inhabitants of Holland have, accordingly, always largely relied upon inland navigation for their means of transport; and their waterways have been developed in proportion to the increased demands of navigation; whilst Amsterdam and Rotterdam are intersected by canals which are connected with the various canals and rivers of the kingdom.

**Waterways of Holland.** The canals are most numerous in the provinces of Friesland and Groningen to the north-east, and in the provinces of North and South Holland and Utrecht, where various canals converge to Amsterdam, which is connected by water with the northern districts by the North Holland Canal and some smaller canals, with the North Sea by the Amsterdam Ship-Canal, and with the Hague, Utrecht, the Maas, the Leck, the Yssel, and the Merwede by canals, and thereby with the other waterways of the country [1]. The canals outnumber the railways in the very flat districts in the north-

---

[1] 'Les Voies de Navigation dans le Royaume des Pays-Bas.' The Hague, 1890, plate 1.

east, and round Amsterdam down to the Hague and Utrecht, where they frequently serve to discharge the drainage waters, and where locks are often only required to separate the different drainage districts. In the eastern and south-eastern parts, however, of the country, where the land gradually rises, and locks are necessary to provide for changes of level, railways have been more developed than canals.

The largest canals of Holland, which serve both for maritime and inland navigation, are the Amsterdam Ship-Canal (Plate 12, Figs. 1 to 6), and the North Holland Canal (Plate 11, Fig. 24), both of which were constructed to afford Amsterdam deep-water access to the sea ; the Voorne, South Beveland, and Walcheren canals, connecting adjacent outlet channels of the Maas and Scheldt ; the Terneuzen-Ghent Canal, affording Ghent a direct access to the estuary of the Scheldt (Plate 11, Fig. 23), and the Eems Canal, connecting Groningen with the sea. The canals of the second class connect the towns with the principal villages ; whilst the Zuider Zee provides communication between the canals of the eastern and western provinces. Smaller canals for local navigation form a third class, comprising mainly small canalized rivers and drainage canals, and also canals constructed for obtaining the peat in the north-eastern districts.

The principal navigable rivers of Holland are the various branches of the Rhine and the Maas, of which the principal are the Hollandsche, the Lower Rhine, Leck, and Maas, the Waal and the Merwede, and the Hollandsch Diep, traversing the southern part of the country in a fairly parallel direction from east to west; whilst the Yssel, branching off near Arnhem, goes northward, giving the north-eastern canals access to the Rhine, and falls into the Zuider Zee. These rivers, moreover, and the waterways in connection with them, afford access to Germany by the Rhine, and to Belgium by the South Willems Canal.

**Lengths of Dutch Waterways.** The length of the rivers

buoyed for navigation in Holland amounts to 350 miles; the length of canals and canalized rivers reaches 2,050 miles; and there are in addition 124 miles of other navigable waterways[1]. This, accordingly, gives a total length of 2,524 miles of inland navigable waterways in Holland. As the country has an area of 12,630 square miles, there are 20 miles of waterways per 100 square miles in Holland, as compared with 6 miles in England, and 3½ miles in France.

**Control of Dutch Waterways.** The principal rivers are taken care of and have been gradually regulated and deepened by the State. Some of the canals are maintained by the State; but the larger number are under the charge of the different provinces; whilst in the several drainage districts, the communes, companies, or individuals, own and maintain the canals. Tolls are levied on vessels navigating the canals, at rates approved by the Government.

**Sizes of Dutch Waterways.** The canals of Holland, having been constructed at various times, and belonging to different bodies, are naturally not uniform in dimensions; but vessels 98 feet long, 18 feet wide, and drawing 6 feet of water can traverse most of the canals. This is the type of vessel, with a length sometimes reaching 131 feet, which is used for general inland navigation throughout Holland, as vessels of larger draught could not navigate the Zuider Zee. The main rivers, however, and some of the canals are accessible to the boats of the Rhine; and these canals have been enlarged from time to time, to provide for the increase in size of these boats, which has been very considerable in the last fifty years in proportion as the rivers have been improved. As the increase in depth obtainable by regulation works on rivers is limited, owing to the reduction of the flow in the summer (p. 59), the low-water navigable channel has been straightened so as to accommodate longer and broader vessels. The largest

[1] L'État et l'Exploitation des Voies Navigables des Pays-Bas.' Ph. W. Van der Sleyden, Manchester Inland Navigation Congress, 1890, p. 2.

vessels passing between Rotterdam and Germany, along the main branch of the Rhine, are 259 feet long, 62¾ feet wide, and 7 feet 10½ inches draught, and have a tonnage of 1,300 tons; but the other branches cannot admit equally large vessels.

The most recent inland canal of importance constructed in Holland is the canal from Amsterdam to the Merwede, which it enters at Gorinchem, completed in 1893 at a cost of about £1,666,000; and it may therefore be regarded as the typical size for a main inland waterway in Holland at the present day[1]. This canal is 42½ miles long, having four reaches, and crossing the Leck at Vianen; it has a depth of 10½ feet, a bottom width of 65⅔ feet, and side slopes of 2 to 1; and the locks have an entrance width of 39⅓ feet, a width of chamber of 82 feet, and an available length of 393¾ feet. The largest vessels that are admitted into the canal are 279 feet long, 34½ feet wide, and 8½ feet draught, somewhat larger than the ordinary large Rhine boats; and the available headway is 21⅓ feet. The navigation along this canal is free of toll; and a considerable traffic has passed through it since its opening in 1893.

Another typical section of the larger Dutch inland canals is the canal from Rotterdam, by the Hague and Leyden, to Haarlem, with a bottom width of 39⅓ to 52½ feet, a depth of 9⅓ feet, and side slopes of 2 to 1; whilst the canals in the peat districts have in general a bottom width of 23 to 26¼ feet, a depth of 5 to 6½ feet, and side slopes of 2 to 1.

The larger canals intended to accommodate sea-going vessels, omitting the Amsterdam Ship-Canal, are from 26 to 55¾ feet wide at the bottom, and 14¾ to 21⅓ feet deep, with side slopes of 2 and 2½ to 1.

**Remarks on Inland Navigation in Holland.** Inland navigation occupies a more important position in Holland than in any other country in the world, owing to the abundance of

---

[1] 'Le Canal reliant Amsterdam à la rivière la Merwede.' P. H. Kemper. Guide du VIᵐᵉ Congrès International de Navigation intérieure. The Hague, 1894, p. 21 and plate.

water, the flatness of the land, and the agricultural nature of the country. In fact, in some low-lying districts, punts have always served the purposes of carts; whilst in Amsterdam and Rotterdam, canals are used, like the streets in other towns, for bulky traffic. The traffic is little restricted by locks, which are sometimes more than 25 miles apart ; and in many cases they are merely introduced to regulate the water-level, and for this purpose they are sometimes provided with reverse gates, which are also adopted to exclude the tide when a lock opens into a tidal river. The improvement of the rivers for navigation, by dredging and training works, has also materially facilitated the discharge of the drainage waters, which is a great advantage for the low-lying lands.

The traffic on the principal waterways has exhibited considerable fluctuations in the last few years ; but a decided increase in traffic has been manifested on the undivided Rhine, the Leck, the canal from Groningen to Lemmer, the South Willems Canal, and the Maastricht and Liége Canal, as well as on the maritime waterways of the Maas to Rotterdam (p. 224), the Amsterdam Ship-Canal, the Terneuzen-Ghent Canal, and the South Beveland Canal. There is also a very considerable traffic on the Zuider Zee, between the inland navigation ports round its shores. The Walcheren Canal between Veere and Flushing has never had a large traffic, having been merely constructed to maintain the communication between the outlet channels of the Scheldt on each side of Walcheren Island, on the barring of the Sloe branch by the Flushing Railway, just as the construction of the South Beveland Canal was necessitated by the barring of the East Scheldt by the same railway. The traffic on the Voorne Canal (p. 222), which was considerable so long as it formed a link in the deepest waterway between Rotterdam and the sea, has almost ceased since the formation of the new deep outlet from the Scheur branch of the Maas, which has so greatly increased the commercial importance of Rotterdam.

INLAND NAVIGATION IN BELGIUM.

Portions of Belgium for some distance inland from the North Sea, and along near the Dutch frontier, are very flat; and though Belgium is not intersected by rivers like the south part of Holland, yet the Scheldt, the Meuse, the Sambre, and other rivers afford important facilities for inland navigation. Belgium, moreover, possesses important coalfields round Mons, Charleroi, and Namur, and great iron and mineral industries between Liége and Namur; whilst Antwerp has risen to the position of one of the great seaports of Europe. Accordingly, although Belgium does not possess natural advantages for inland navigation to the same extent as Holland, nevertheless, owing to the inland position of its capital, the proximity of its coal and mineral districts to France, and the deep-water access to the sea provided for Antwerp by the Scheldt, its inland navigation has been largely developed.

**Waterways of Belgium.** The main outlets for the navigation to the sea are by the Scheldt from Antwerp, and by the Ghent-Terneuzen Canal. Ghent is also connected with Termonde and Antwerp by the Scheldt, and by canal with Bruges and Ostend; and Ostend is in communication by canals with Nieuport and Dunkirk. A network, moreover, of canals and canalized rivers connects Antwerp, Ghent, and Mons with the northern waterways of France; whilst Brussels is joined to the Scheldt and Antwerp by the Brussels and Rupel Canal (opened first in 1561) and the river Rupel, and with Charleroi, Namur, and Liége, by the Charleroi and Brussels Canal and the canalized Sambre and Meuse. Antwerp is also connected with Maastricht and Liége by canals skirting the Dutch frontier, and joining the Meuse at Liége[1]. In this manner the principal towns of Belgium are in communication by water with each other, and with France and Holland. Two connecting links, which are in course of construction, are

---

[1] 'Guide du Batelier.' Ministère des Travaux Publics, Brussels, 1889, Maps.

the Canal du Centre for uniting the Mons and Condé Canal with the Charleroi and Brussels Canal, and consequently Mons with Brussels. Namur, and Liége; and the Ypres and Lys Canal, joining the Yser to the Lys, and therefore to the Scheldt. The waterway connecting Antwerp with Dunkirk, by Ghent, Bruges, and Ostend, passes through such flat low-lying country that, in its whole length of 118¾ miles, the greatest variation in its level is only 10 feet; whilst the waterway from Antwerp, skirting the Dutch frontier to the east, and passing by Maastricht, Liége, and Namur, to the French frontier along the Meuse, rising 314 feet in a total length of 168 miles, has a reach, terminating at Maastricht, 43½ miles long without change of level. The canal, however, connecting the Rupel with the Sambre, passing by Brussels and Seneffe to Charleroi, has to rise 353 feet, by forty-four locks, in a distance of 31 miles between Brussels and the summit tunnel at Seneffe piercing the ridge separating the basins of the Scheldt and the Meuse; and it descends again 70 feet by eleven locks to Charleroi, in a distance of 9½ miles. A steeper ascent, moreover, has to be accomplished by the Canal du Centre, in course of construction, which rises 294 feet in 13 miles between Mons and La Louvière, the greater part of the ascent being effected by the lifts previously referred to (p. 408).

**Lengths of Belgian Waterways.** The total length of waterways in Belgium on which navigation is carried on, amounts to 1,370 miles; but as 123 miles of this consist of merely floatable rivers, the actual length of navigable waterways is 1,247 miles. This length is composed of 303 miles of open navigable rivers, 347 miles of canalized rivers, 454 miles of large canals, and 143 miles of small canals[1]. The State controls 1,118 miles of these waterways; and the greater part of the remainder belongs to the provinces, communes, and drainage boards, only 59 miles of canalized rivers and large canals having been conceded to companies. Belgium

[1] 'Guide du Batelier, Annexe II, Modifications et Compléments, 1893,' p. 89.

has an area of 11,373 square miles, and therefore it possesses nearly 11 miles of navigable waterways per 100 square miles, only a little more than half the proportion possessed by Holland, but nearly double that of England, and three times that of France.

**Sizes of Belgian Waterways.** The Scheldt affords a depth of 26¼ feet at low water up to Antwerp; and there is a minimum depth of 14¾ feet up to the mouth of the Rupel, which is reduced to 6¼ feet beyond Termonde up to Ghent, which is the tidal limit. Besides, however, a depth at high water of 17 feet by the Scheldt, Ghent possesses a much shorter route to the sea by the Ghent-Terneuzen Canal, with a minimum depth of nearly 20 feet. The largest regular inland canals are the canal from Ostend to Bruges, with a depth of 13½ feet, and locks 197 feet long and 27 feet wide; the canal from the Rupel to Brussels, with a depth of 10½ feet, and locks 128 to 249 feet long and 24⅜ feet wide; and the canal from the Rupel to Louvain, with a depth of nearly 12 feet, and locks 180 feet long and 27 feet wide.

Most of the other main waterways have depths of 6½ to 7⅕ feet, and locks of somewhat variable dimensions, but generally not less than 131 feet in length and 17 feet in width. The canal, however, from Brussels to Charleroi, with a depth of 6½ feet up to its summit-level at Seneffe, has locks along this portion only 62 feet long and 8¾ feet wide, and can only admit vessels of 70 tons; whereas the Canal du Centre, which will join it near Seneffe, has been given a depth of nearly 8 feet, with locks and lifts 134 to 141 feet long and 17 to 17¾ feet wide, in order to admit vessels of 300 to 400 tons. The restricted section of the canal between Brussels and Seneffe constitutes a serious impediment to navigation along this important link; but it is proposed to enlarge it. The large Rhine and Dutch vessels can only get up to Liége, Brussels, and Ghent.

**Traffic on Belgian Waterways.** The largest inland traffic, exceeding an average of one million tons in the year, is confined to the Scheldt between Antwerp and the Rupel, up this river and by canal to Brussels, and along the Ghent-Terneuzen Canal, with averages of 1,880,000, 1,439,000, 1,120,000, and 1,073,000 tons respectively, in 1893[1]. The traffic on the Scheldt between the Rupel and Termonde is only slightly less, with an average of 892,000 tons in 1893, and on the Meuse between Namur and the French frontier, with an average of 823,000 tons. The average traffic by water is about three-quarters of a million tons between Namur and Charleroi, and from Maastricht to the northern frontier; and about two-thirds of a million tons between Charleroi and the French frontier, Liége and Maastricht, and Antwerp and the Maastricht Canal; whilst the traffic on the Charleroi and Brussels Canal, the Dendre, and the Upper Scheldt, is under half-a-million tons. There is very little traffic along the canal between Ostend and Bruges in spite of its ample dimensions, and practically no traffic along the waterways in the western corner of Belgium. Accordingly, it appears far more important to improve the waterways between Antwerp and Brussels, and from Antwerp round by Maastricht to Liége, Namur, and the frontier, and especially the canal from Charleroi to Brussels, than to construct the proposed ship-canal to Bruges.

Antwerp, Brussels, and Ghent, and the coal and mineral districts are the sources of the traffic by water, for the waterways connecting these towns and districts have a large traffic, notwithstanding the ample railway accommodation which Belgium possesses; whilst the number of locks, and the inadequate size of the Charleroi and Brussels Canal, have only prevented the proper development of the traffic, without driving it away. On the other hand, the facilities for navigation afforded by the level canal from Ghent to Ostend, and by

[1] Carte figurative du Mouvement des Transports sur les Voies Navigables de la Belgique en 1893.' Ministère des Travaux Publics, Brussels.

the canals in the low-lying lands near the coast, have proved useless in the absence of bulky traffic suited for waterways. The coal traffic predominates on the Sambre and Meuse between Namur and the French frontier, and on the Charleroi and Brussels and the Mons and Condé canals; the traffic in coal and minerals is about equal on the Meuse between Namur and Liége; whilst minerals, metals, and building materials constitute the main traffic from Liége round by Maastricht to Antwerp, and on the Blaton and Ath Canal, the Dendre, the Upper Scheldt, and the Lower Scheldt between Termonde and Ghent. Between 1881 and 1893, the traffic was trebled on the Meuse between Liége and the French frontier; it was about doubled on the Ghent-Terneuzen Canal, the Sambre, and the Lys; and it increased considerably on the Lower Scheldt, and on the canal between Antwerp and the Maastricht Canal.

Tolls are levied on vessels navigating the canals and canalized rivers, in proportion to the weight of cargo carried, and the distance traversed; and the tolls on canals are about three times the rate of the tolls on canalized rivers.

**Average Traffic on Belgian Railways and Waterways compared.** The State railways of Belgium, with a traffic of 750,000,000 ton-miles in the year, and a length of 1,607 miles, have an average traffic of 466,000 tons. Considering now only the main waterways in Belgium, having a length of 937 miles, and a yearly traffic of 359,000,000 ton-miles, the average traffic on them is 383,000 tons, only about one-sixth less than the traffic on the railways[1]. This result shows that, under favourable conditions as regards the nature of the traffic, with some good rivers, and a country for the most part fairly suitable for canals, it is possible for the principal inland waterways to obtain a traffic about equal to that of railways,

---

[1] 'Données Statistiques sur la Situation, l'Exploitation, et les Dépenses de Construction et d'Entretien des Voies Navigables de la Belgique.' A. Dufourny, Manchester Inland Navigation Congress, 1890, p. 11.

notwithstanding the greatly superior traffic arrangements on railways than on most waterways.

## INLAND NAVIGATION IN SPAIN.

Spain, owing to its mountainous character and high table-lands, is not well suited for inland navigation, except along the lower parts of some of its largest rivers; and its canals have been mostly constructed for irrigation, to which reference has been previously made (p. 450).

**Navigable Rivers of Spain.** The rivers of Spain are torrential in character, and therefore are not naturally adapted for navigation, except for comparatively short distances from their outlets; and little has been done for the most part to improve their condition. Some canalization works for rendering the Ebro accessible to steamers up to Saragossa, were carried out in 1851–58 as far as Escatron; but the improvement of the river between Escatron and Saragossa was abandoned as impracticable; and the construction of railways in the neighbourhood deprived the river of most of its traffic, and led to the utilization of its waters for irrigation[1]. The San Carlos Canal also, $6\frac{3}{8}$ miles long, constructed to avoid the delta of the Ebro, having become very shallow by the deposit of alluvium, is now used for irrigation; and the traffic on the river has fallen very low.

The Guadalquivir has been rectified and trained in some places between Seville and the sea, whereby the distance has been reduced to 54 miles, and the minimum depth increased to 17 feet; and it possesses a fair traffic, which however appears to be more maritime and coasting than strictly inland river traffic. The Guadiana also is navigable for large vessels for 41 miles from its mouth, and has some inland traffic as well as a maritime trade. The Tagus and the Douro are both navigable in Portugal; but the Tagus would require

[1] 'La Navigation intérieure en Espagne.' André de Llaurado, Manchester Inland Navigation Congress, 1890.

considerable works to render it properly navigable in Spain ; and the competition of the railways has driven away the traffic from the lower portion of the Spanish part of the Douro, which is impeded by shoals and rapids.

**Spanish Inland Canals.** Two canals only have been constructed in Spain for inland navigation, namely the Castile and Aragon canals. The Castile Canal goes from Alar del Rey to Valladolid, with a branch from Serron to Rio Seco, having a total length of 130 miles. The works were partially carried out between 1753 and the close of the century; they were recommenced in 1831 ; and the canal was completed in 1849. The canal has forty-nine locks, with lifts averaging $9\frac{3}{4}$ to 11 feet ; its depth of water is $6\frac{1}{2}$ feet ; and it is used by vessels of 34 tons. The adjacent railways have diverted a large part of the traffic, which has fallen to an average of about 22,000 tons per annum.

The Aragon Canal, going from Tudela to Saragossa, $54\frac{3}{4}$ miles long, constructed in the sixteenth century for irrigation (p. 451), was enlarged and rendered navigable in 1770–90, and is accessible to vessels of 100 tons drawing $6\frac{1}{2}$ feet of water. The opening, however, of the railway from Saragossa to Pampeluna, deprived the canal of all its traffic, except very bulky goods ; and the chief value of the canal now consists in supplying water for irrigation and water-power.

## INLAND NAVIGATION IN ITALY.

Italy, with its great extent of sea-coast and numerous harbours, its narrow width, its mountainous regions to the north and north-west, and the chain of the Apennines running down its whole length, is not adapted by nature for inland navigation, except in the plains of Lombardy from Milan to the shores of the Adriatic, along the lower portions of the valleys of the Arno and the Tiber, and in the Pontine marshes [1].

[1] ' Relazione sui Lavori del Congresso di Navigazione Interna tenuto a Manchester nel 1890, dei Delegati del Ministero dei Lavori Pubblici.' G. Bompiani and L. Luiggi, Giornale del Genio Civile, Parte non Ufficiale, Rome, 1891, p. 85, and plate 2.

**Development of Inland Navigation in Italy.** The natural facilities for navigation afforded by the Italian lakes and the river Po, were developed during the middle ages by the formation of canals serving the double purpose of navigation and irrigation. The earliest of these canals was the Naviglio Grande, constructed in the twelfth century (p. 443), connecting Milan with the Ticino, and consequently with Lake Maggiore; and Milan was subsequently connected more directly with the Po by the Pavia Canal. A few years later, the Modena, Bologna, and Padua canals were constructed on the eastern side; and subsequently the Bereguardo, Martesana, and Padermo canals were carried out in the neighbourhood of Milan, as well as various canals in the Venetian provinces [1]. Other canals were formed at various periods, the system of canals being most developed in the lower basin of the Po below Ostiglia, and inland from Venice; whilst the rivers Brenta, Sile, Piave, Livenza, and Tagliamento have been connected near their outlets by canals bordering the Adriatic coast. Ravenna also has been connected with the Adriatic by the Corsini Canal; and the shallow mouth of the Tiber has been avoided by the construction of the lateral Fiumicino Canal, branching off from the Tiber at Capo due Rami. Three canals in the lower valley of the Arno connect Pisa, Ripafratta, and Bientina with the port of Leghorn; whilst the Pontine Canals, though available for navigation, serve almost wholly for the drainage of the Pontine marshes.

Works for the improvement of the navigation on the Po, some of its tributaries, the Venetian rivers, and some of the canals, are being gradually carried out; and extensive works are in progress for improving the Tiber, especially through Rome, mainly for the prevention of floods; but no works of importance for the extension of inland navigation have been executed within recent times.

[1] Management and Expenses of Working of Inland Navigation in Italy. G. Bompiani and L. Luiggi, Manchester Inland Navigation Congress, 1890.

**Lengths and Sizes of Italian Waterways.** The total length of the navigable portions of the rivers of Italy is 1,186 miles, and of the canals 549 miles, exclusive of the Pontine drainage canals, which have a length of 113 miles. The Po from a little above Piacenza to its mouth, a distance of 230 miles, has a minimum depth of 8½ feet; and the lower extremities of some of its tributaries, and of some of the Venetian rivers, have minimum depths of 10 to 16½ feet; and these, as well as the Tiber below Rome to the entrance of the Fiumicino Canal, belong to the first class of navigable rivers.

The canals having been constructed at various periods, do not possess that uniformity in dimensions so important for facilitating the development of inland navigation. They are divided into three classes; and except in the Pontine marshes, more canals belong to the second class than to the other two. The largest canals are the Corsini Canal from Ravenna to the sea, nearly 7 miles long, having a depth of 13 feet, and capable of accommodating vessels of 300 tons; the Fiumicino Canal, 3 miles long, with a depth of 8½ feet, and navigated by vessels of 200 tons; and the outlet channel from the Pontine drainage canals, 3¾ miles long, and 11½ feet minimum depth though narrower than the other two, but having hardly any traffic on it. The other canals of the first class, having a minimum depth of 6½ feet, are the series of short canals connecting the Venetian rivers north-east of Venice, and four canals leading to the Venetian lagoons; and their ordinary bottom width is 16½ to 19½ feet. The remaining canals have depths ranging for the most part between 3¼ and 5 feet, and bottom widths comprised generally between 13 and 33 feet.

Owing to the flatness of the country round Venice and in the lower valley of the Po, the canals of these districts have generally no locks; and with the exception of the Brenta navigation, which has four locks, none of them have more than one or two locks, ranging in width from 16½ to 33 feet, and in length from 51 to 160 feet. The only canals with several locks are the Pavia with twelve locks, the Bereguardo with thirteen, and the Bologna Canal with ten. Four of the canals in the lower valley of the Po can admit vessels of 150 tons, and most of the Venetian canals can pass vessels of 100 tons ; vessels of 30 to 90 tons navigate several of the canals according to their sizes ; but on two of the canals of the Po valley, and on the canals of Tuscany the largest barges carry only 20 tons.

**Control of the Italian Waterways.** All the waterways of Italy, with the exception of the Pontine drainage canals, are managed, maintained, and improved by the State. In 1879 the tolls were removed from all the navigable waterways of Italy, which has caused a considerable increase in the inland navigation. The want of uniformity, however, in the sizes of the network of waterways has impeded this growth of traffic.

**Traffic on Italian Waterways.** In the middle ages, the waterways were the great means of communication in Northern Italy ; and they retained a considerable commercial importance till the advent of railways caused a large decrease in the traffic by water. Inland navigation, however, has revived again in Italy, as in other parts of Europe, owing to the advantages which it presents for the conveyance of agricultural produce, building materials, and other bulky goods. This improvement in traffic, which is specially noticeable in the basin of the Po and on the Tiber, is due to improved commercial conditions and a general increase in the traffic of the country, as well as the attractions afforded by the abolition of tolls.

The Valle Canal near the Adriatic, to the south of the Venetian lagoons, has the largest average yearly traffic, reaching one million tons; next comes the Po, with three-quarters of a million tons ; and then follow, at a considerable interval, the Brenta navigation, with about 310,000 tons yearly, the Dolce Canal and the river Sile, with an annual traffic of about a quarter of a million tons, and the Lorco Canal with 210,000 tons. Then follow in order the Cagnola and Pontelongo canals, Lake Maggiore, the Ticino, the Adige, and the Cavanella di Po and the Pisa and Leghorn canals, with a traffic of between 200,000 and 150,000 tons ; whilst the traffic on the Lake of Como, the Mincio, the Arno, the Tiber, and the Sile-Piave, Pavia, Fiumicino, Corsini, and Naviglio Grande canals exceeds 100,000 tons per annum.

Accordingly, although navigation in Italy is confined to certain suitable districts, and the waterways have not been improved so as to keep abreast of the growing requirements of trade, there is a very fair traffic on some of the waterways, which might doubtless be considerably augmented by making the main lines in the basin of the Po and round Venice uniform in dimensions.

# CHAPTER XXI.

## INLAND NAVIGATION (*continued*).

Conditions affecting Inland Navigation. *Inland Navigation in Sweden :*—Principal Waterways connecting Lakes; Traffic. *Inland Navigation in Germany :*—Navigable Rivers, lengths and depths, sizes of locks; Inland Canals, lengths, sizes of locks; Lengths of Waterways; Traffic on German Waterways. *Inland Navigation in Austria-Hungary :*—Austrian Waterways, lengths, traffic; Hungarian Navigable Rivers; Canals; Transport, sizes of vessels, and nature of trade; Traffic on Hungarian Waterways and Railways compared. *Inland Navigation in Russia :*—Navigable Rivers; Inland Canals connecting Rivers, Caspian connected with Baltic, Marie Navigation, Baltic connected with Black Sea, White Sea connected with Baltic and Caspian; Recent Works; Lengths of Russian Waterways; Traffic. *Inland Navigation in India :*—Limits; Navigable Rivers; Navigable Canals, combined with Irrigation, constructed for Navigation. *Inland Navigation in Canada :*—River St. Lawrence, and the Lakes; St. Lawrence Canals; Welland, Sault-Sainte-Marie, and other Canals; Depth of Canals, and Sizes of Locks; Traffic. *Inland Navigation in the United States :*—Lakes and Rivers; Mississippi and Tributaries; Hudson and minor Rivers; Erie, Champlain, Morris, Illinois and Michigan, and Chesapeake and Ohio Canals; Central and Southern Transportation Routes; St. Mary's Falls, and St. Clair Flats Canals; Hennepin Canal; Chicago Drainage Canal, objects and route; Sizes of Locks on Waterways; Sections of Canals; Traffic on State of New York Canals compared with Railways; Traffic on the principal Rivers of the United States. Concluding Remarks, control of rivers, and extension of navigation.

INLAND navigation has been shown in the preceding chapter to depend greatly upon the capabilities of the natural waterways, the distance from the sea-coast, and the general physical conditions of a country, as well as upon the nature of the traffic to be accommodated. Its development, however, also depends upon the density of the population, which creates an increased demand for every available means of intercommunication. Thus Holland possesses the largest extent of waterways in proportion to its area, on account of its exceptionally favourable physical conditions; whilst Belgium comes next, amongst the countries of Europe, owing to its mineral resources and the density of its population. The other European countries, however, do not present a similar development of navigable waterways in comparison to their areas; for their inland navigation is, for the most part, confined to special districts, where the conditions of the country, or of the traffic are particularly favourable, or to natural waterways extending a long distance inland. Large portions of northern, central and southern Europe, are too mountainous to render inland navigation a convenient method of communication; whilst Russia, though possessing a considerable extent of flat plains far removed from the sea-coast, has too scattered a population to render costly works for a large development of inland waterways expedient at the present time.

## INLAND NAVIGATION IN SWEDEN.

Norway is too mountainous throughout to be suitable for inland navigation, though its innumerable fiords, stretching considerable distances inland, afford it some compensation in respect of maritime facilities for internal communication. Sweden also is unsuited for inland navigation, except in its southern part, which is flat and has some large inland lakes.

**Waterways of Sweden.** Though canals were commenced in Sweden at the beginning of the fifteenth century, the first canal with locks, giving the town of Eskilstuna navigable access to Lake Mälar, was constructed in 1596–1606. This canal also, together with the river Eskilstuna, connects Lake Hjelmar with Lake Mälar. The most important of the canals unite various lakes together, aided generally by the canalization of the rivers draining the lakes; and sometimes short lateral canals have been formed to avoid the rapids of the rivers.

There are about thirty systems of artificial waterways in southern Sweden; and the most important of these are the Göta and Trollhätta navigations, commenced in the sixteenth and seventeenth centuries, but completed and enlarged in the earlier part of the nineteenth century, which utilizing as far as possible the rivers, whilst avoiding the rapids, unite several lakes in their course, and form a continuous navigation between the Baltic and the Cattegat and consequently to the North Sea, 257 miles long[1]. The East Göta navigation, starting from the Baltic 70 miles south of Stockholm, rises by locks to Lake Wetter, passing through some smaller lakes in its course; and the West Göta navigation connecting the large lakes Wetter and Wener, rises to Lake Viker the summit-level of the waterway, 298½ feet above the Baltic, which rise is surmounted by thirty-nine locks. The West Göta navigation then descends 158¾ feet to Lake Wener, by nineteen locks; and Lake Wener is connected with the Cattegat at Gothenburg by the Trollhätta Canal, which descends the 143¾ feet that the level of the lake is above the sea-level, by sixteen locks. The locks are 117 feet long and 24⅛ feet wide, and have 9¾ feet depth of water over their sills. The branch canals Kinda and Dalsland join the main waterway at the lakes Roxen and Wener respectively. The Kinda Canal, constructed in 1865–71, proceeds to the south, and by the canalization of the river Stanga penetrates 49¾ miles into the interior, connecting the lakes Rengen and Jernlunden with Lake Roxen, and rising 171½ feet by fifteen locks to an elevation of 277 feet. The Dalsland Canal, formed in 1865–69, unites a series of lakes with Lake Wener, and rising 193½ feet by twenty-five locks to an altitude of 337 feet above sea-level, provides a navigable waterway, 158 miles long, penetrating as far as Norway.

Lake Mälar is connected with the Baltic at Stockholm by the help of a lock, through which about 14,000 vessels pass in the year; and another canal unites this lake to the Baltic at the town of Södertelge.

---

[1] 'The Canals of Sweden,' Colonel A. M. Lindgren, Manchester Inland Navigation Congress, 1890.

**Traffic on Swedish Waterways.** The waterways in Sweden with the largest traffic, next to that passing through the lock at Stockholm, are the Trollhätta and Göta navigations, and the Södertelge, Dalsland, Strömsholm, and Kinda canals. The Göta and Trollhätta navigations afford a short route, free from the dangers of the Sound, for vessels trading between Stockholm and Gothenburg; and the principal waterways admit coasting vessels up to 250 tons.

## INLAND NAVIGATION IN GERMANY.

The extensive plains of Northern Germany, stretching from the Dutch frontier to Memel, and traversed by the Rhine, the Ems, the Weser, the Elbe, the Oder, the Vistula, and the Niemen, offer great facilities for inland navigation, which are extended along the larger river basins, such as the Rhine, the Elbe, and the Vistula, into the hilly districts to the south.

**German Navigable Rivers.** The chief rivers of Germany, possessing large basins, and a moderate fall in the lower portions of their course, have been rendered navigable by regulation works at a moderate cost; and they form the principal inland waterways of the country, as they are mostly navigable for long distances into the interior, though with diminished available depths. Thus the Rhine, which provides a waterway for the large Rhine boats up to Mannheim, 352 miles from Rotterdam, with a general average depth of between $12\frac{1}{2}$ feet and $7\frac{1}{2}$ feet up to Karlsruhe[1] (p. 60), is more or less navigable up to Bale, 532 miles from the sea; though for the last 37 miles the average depth is only $3\frac{1}{4}$ feet, and the minimum depth at the lowest stage of the river is half that amount[2]. The Weser, which is accessible for sea-going vessels up to Bremen. is navigable up to its termination at Münden, 271 miles from Bremerhaven, with average depths decreasing

---

[1] The average depth given is the available depth at the average level of the river; and the minimum depth is the available depth when the river is at its lowest level.

[2] 'Karte der Deutschen Wasserstrassen,' Sympher and Maschke, Berlin, 1893.

from $8\frac{1}{5}$ feet to $3\frac{1}{2}$ feet, from whence the navigation continues a little further inland along its tributaries the Werra and the Fulda. The Elbe, which enables large sea-going vessels to reach Hamburg, 65 miles from the sea, is available for inland navigation up to the Austrian frontier, 382 miles higher up; but the depth, which averages $8\frac{1}{5}$ feet, with a minimum of 3 feet, just above Hamburg, is reduced to 6 feet, with a minimum of $2\frac{3}{5}$ feet, near the frontier. The Oder, possessing an average depth of $19\frac{2}{3}$ feet up to Stettin, 45 miles from the sea, which is soon reduced above to 10 feet, retains a navigable depth averaging $6\frac{1}{2}$ feet, with a minimum of $3\frac{1}{4}$ feet, from a little below Frankfort on the Oder up to 398 miles from the sea; and it continues navigable up to Ratibor, near the Austrian frontier, with a depth reduced to 5 feet average, and $1\frac{2}{3}$ feet minimum. The Vistula and Niemen are navigable from the Baltic to the frontier, their lengths in Germany being 153 miles, and 102 miles respectively, with an average depth of about $5\frac{1}{2}$ feet. The Danube also is navigable in Germany from Ulm downwards, the distance along it to the frontier being 238 miles, with an average depth of $5\frac{1}{5}$ feet at Ulm, which is increased to $7\frac{1}{5}$ feet before reaching the frontier.

The principal navigable tributaries of some of the above rivers, possessing for the most part an open navigation, are the Moselle, the Main above Frankfort, the Neckar, the Havel, the Netze, the Wartha, and the Inn. The rivers which have been canalized, generally over comparatively short lengths, are the Ems, the Lippe, the Ruhr, the Lahn, the Main up to Frankfort (p. 69, and Plate 4, Figs. 5 and 6), the Elde, the Havel above Fürstenburg, the Spree above Berlin, the Upper Netze, and short lengths of the Saar, the Altmühl, and the Alle. The locks on these rivers vary in size, according to the available channel and importance of the traffic, from about 90 feet long, $15\frac{3}{4}$ feet wide, and $3\frac{1}{2}$ feet depth of water, up to 360 feet long, $31\frac{1}{2}$ feet wide, and 8 to $6\frac{1}{4}$ feet depth of water according to the state of the river, which is the size of the new lock on the Spree, and 1,148 feet long, 34½ feet wide, and 8⅓ feet depth of water for the lengthened locks on the Main.

**Inland Canals in Germany.** Canals occupy quite a subordinate position in Germany in comparison with the river navigations; and they have mostly been constructed to connect the rivers, and thereby extend the lines of communication by water. The principal of these canals are the Ludwigs Canal, 83½ miles long, connecting the Main with the Danube, through the canalized river Altmühl; the Rhine-Marne Canal, and the Rhine-Rhone Canal, only partially in Germany; the Saar Canal, 40⅔ miles long, joining the latter canal; the Finow Canal, 44¾ miles long, from Zehdenick to Hohensaaten, connecting the Havel and the Oder; the Oder-Spree Canal, 54⅔ miles long, providing a waterway between the Elbe and the Oder; and the Oberland Canal. There are also shorter canals connecting the Ems with the Jade, and with the Hunte, the Havel at Plaue with the Elbe, and the Vistula with the Netze, and consequently with the Oder.

These canals have depths of 6½ to 7 feet in the case of the Oder-Spree and Ems-Jade canals, and ranging between 5 and 6 feet in most of the other canals. The locks vary considerably in size on the different canals, from a maximum length of 213 feet on the Plaue Canal, and a maximum width of 28 feet, and depth on the sill of 8⅓ feet on the Oder-Spree Canal, down to a minimum length of 70 feet on the Hadelner Canal, and a minimum width of 13 feet and depth of 4 feet on the Klodnitz Canal; but the lengths, for the most part, range between about 100 and 135 feet, the widths between 15 and 17½ feet, and the depths between 4¼ and 5¼ feet.

A canal has been projected, running east and west through the centre of western Germany, for placing Berlin in direct communication with the Rhine. This canal, starting from the Elbe a little north of Magdeburg, would pass by Hanover, and crossing the Weser near Minden, would turn to the south on reaching the Ems, and following the Ems Canal would eventually enter the Rhine at Ruhrort. This canal would connect western with eastern Germany by water, and would enable the products of the Rhine valley to be carried by water direct to Berlin.

**Lengths of German Waterways.** The rivers used for open inland navigation in Germany have a total length of 4,326 miles, exclusive of the portions accessible to ocean-going vessels. The length of canalized rivers amounts to 774 miles, making a total length of inland river navigation of 5,100 miles. The inland canals extend over 1,221 miles, giving a total length of inland waterways of about 6,321 miles. There is also a prolongation of the Ems Canal to Dortmund in progress, about 84 miles long; and the improvement of about 370 miles of waterways is being carried out. As Germany has an area of 211,168 square miles, it possesses 3 miles of navigable inland waterways per 100 square miles of country. half the proportion of English inland waterways, and somewhat less than that of inland waterways in France (p. 488), where the smaller length of navigable rivers is more than compensated for by the much greater development of canals than in Germany.

**Traffic on German Waterways.** A glance at the map indicating the traffic on German waterways for 1885, shows that traffic on these waterways is very unevenly distributed[1]. Thus the Rhine possessed a traffic, in 1885, of 2,129,000 tons at Mannheim, the second port on the river, increasing to 4,474,000 tons below Ruhrort, the principal river port in Germany. Next comes the Elbe with a traffic at the Austrian frontier of 1,848,000 tons, increasing to 2,616,000 tons at Hamburg, but differing from the Rhine in its down-stream traffic being in excess of the traffic up-stream. Its principal ports next to Hamburg, which has a river traffic exceeded only by Ruhrort and Berlin, are Magdeburg and Dresden.

[1] 'Karte des Verkehrs auf Deutschen Wasserstrassen im Jahre 1885,' Sympher, Berlin, 1889.

The Oder occupies the third position, owing to its connections by water with Berlin through the Finow Canal and the river Havel, and by the Oder-Spree Canal, its inland traffic increasing from 489,000 tons at Breslau to a maximum of 1,229,000 tons below Cüstrin to Hohensaaten, owing to the influx of traffic from the Netze, which by its junction with the Vistula by a canal, draws supplies from Austria by that river. The Vistula and the Niemen had a traffic in 1885 ranging between 458,000 tons at Fordon, near Bromberg, and 760,000 tons at the Austrian frontier, and between 361,000 tons at Memel and 693,000 tons at the Russian frontier, consisting chiefly of down-stream traffic from Austria and Russia to the seaports of Dantzic and Memel. The traffic on the Main between Frankfort and the Rhine has greatly augmented since its canalization (p. 69); and the canalized Saar and Saar Canal have a traffic of 843,000 tons. The traffic, moreover, in 1885, ranged on the Upper Rhine from 364,000 tons above Mannheim to 122,000 tons at Strassburg; on the Neckar, from 280,000 tons at Heilbronn to 363,000 tons at its mouth; on the Main above Frankfort, from 221,000 tons at Würzburg to 345,000 tons at Frankfort; and on the Rhine-Marne Canal, from 334,000 tons as far as the Saar Canal to 612,000 tons from thence to the French frontier. The maximum traffic, however, on the Rhine-Rhone Canal was only 221,000 tons, and on the Weser above Bremen 173,000 tons; whilst the traffic on the remainder of the German waterways is comparatively insignificant.

The traffic on the whole of the German waterways rose from 1,773 million ton-miles in 1875, to 2,935 million ton-miles in 1885, owing to the improvements effected in the principal waterways during that period[1]; and as the traffic on the Rhine, between Strassburg and the Dutch frontier, was 970 million ton-miles in 1885, and on the Elbe down to Hamburg

---

[1] 'Der Verkehr auf Deutschen Wasserstrassen in den Jahren 1875 und 1885,' Sympher, Manchester Inland Navigation Congress, 1890.

794 million ton-miles, the traffic on these two rivers alone amounted to three-fifths of the whole traffic. The traffic, moreover, on the large waterways increases with their accessibility; whilst the traffic on the smaller waterways has not grown. The average traffic on the whole of the waterways was 472,000 tons in 1885; whilst the average traffic on the German railways was only 443,000 tons, the average journey by rail however, being, only 103 miles as compared with 217 miles on the waterways. Owing to the much greater length of the railways, amounting to about 23,000 miles, the goods traffic on the railways amounts to about three times the traffic on the waterways.

### Inland Navigation in Austria-Hungary.

Austria is in a special degree unfavourably situated for inland navigation, as it possesses no large river basins stretching down to its very limited sea-coast. Several large rivers, indeed, have their sources amongst the mountains of Austria, as for instance the Elbe, the Oder, the Vistula, and the Dniester, but they do not become important waterways till after they have passed the frontier; whilst the Danube, which, rising in Germany, is the principal river of Austria, enters Hungary a few miles below Vienna, at a long distance from its outlet.

**Austrian Waterways.** The Danube, the main navigable waterway of Austria, connects Vienna by water with Germany, and by canal with the Main and the Rhine on the one side, and with Buda-Pesth, and eventually with the sea on the other side. The Danube and its tributaries the Inn, the Traun, and the Enns, have a total navigable length in Austria of 382 miles; but the tributaries serve mainly for floating down timber from the forests for supplying Vienna; and the length of the Danube alone in Austrian territory is 227 miles [1].

---

[1] 'Rôles respectifs des Voies Navigables et des Chemins de Fer dans l'Industrie des Transports en Autriche.' A. Schromm, Congrès international de Navigation intérieure: Paris, 1892, p. 15.

The Elbe is navigable in Austria for 68 miles, and with its tributary the Moldau affords a waterway 218 miles long. The Vistula, with three tributaries, furnishes a navigable length of 255 miles ; but the traffic is wholly down-stream.

The hilly character of Austria is unfavourable for the formation of artificial waterways; but a canal has been constructed between Vienna and Neustadt, 40 miles in length. Railways, accordingly, which extend over a length of 9,500 miles in Austria, afford means of communication better suited to the character of the country than canals.

**Traffic on Austrian Waterways.** The Elbe possesses the largest traffic, of about 676,000 ton-miles in the year, owing to the facility with which coals from the extensive coalfields in the neighbourhood of the river are conveyed down-stream into Germany, along a portion of the river which has been enabled by regulation works to accommodate vessels of 500 tons. The traffic on the Danube, which consists mainly of wood for fuel and timber down-stream, and wheat up-stream, amounts to about 373,000 ton-miles ; but the conveyance of wheat, which constitutes about a fourth of the whole traffic in a good season, is hampered by the lowness of the river generally at the time of harvest, which prevents the vessels being fully loaded. The traffic on the Moldau, which is mainly timber floated down, amounts to about 191,000 ton-miles in the year; whilst the traffic on the Vistula, the bulk of whose trade consists of timber floated down and coal, is only about 21,000 ton-miles.

**Hungarian Navigable Rivers.** The comparative flatness of a considerable portion of Hungary renders it far more suitable for inland navigation than Austria ; and, moreover, the Danube, in flowing through Hungary, is fed by several large tributaries, some of which are navigable for considerable distances [1], as for instance the Theiss for 304 miles, the Save for 375 miles, and the Drave for 142 miles. The Danube is more or less

[1] 'Les Voies Navigables de la Hongrie,' Buda-Pesth, 1892, p. 11, and plate i.

navigable throughout almost the whole of its course through Hungary, along a length of 604 miles, and is being gradually improved by regulation works ; and the remainder of the river is floatable. Its principal tributaries also are floatable for long distances; and the timber felled in the northern forests and Transylvania are conveyed to the Danube in this way.

**Inland Canals in Hungary.** Though a considerable portion of Hungary is sufficiently flat for the construction of canals at a reasonable cost, and several canals have been proposed for improving the communications by water, and for connecting existing waterways, so as to shorten the routes, only two navigable canals have been constructed hitherto in Hungary, namely the Bega Canal joining Temesvar to Titel on the Theiss near its confluence with the Danube, and the Franz Canal, with its branches, connecting Baja with Ujvidek and Foldvar, and the Danube more directly with the Theiss. The Bega Canal, originally formed by the Romans, and reconstructed in the eighteenth century, is much hampered by scarcity of water in the dry season, and by the floods of the Bega, which pass down it; but, nevertheless, it possesses a fair traffic. The Franz Canal was originally opened in 1802, but it was subsequently extended from 73 miles to 146 miles, by works completed in 1875; and it serves for irrigation as well as navigation. The traffic on the extended canal rose steadily up to 1879, but since 1885 it has distinctly declined.

**Lengths of Hungarian Waterways.** The total length of navigable rivers in Hungary amounts to 1,653 miles; the two canals together extend over 217 miles ; and Lake Balaton affords a navigation of 21 miles. Accordingly, the navigable waterways of Hungary are 1,891 miles in length; and in addition most of the tributaries of the Danube are floatable for a total length of about 1,180 miles.

**Transport on Hungarian Waterways.** The vessels navigating the rivers of Hungary vary considerably in dimensions and draught. Thus the steamboats of the Danube Navigation Company are from 74 to 250 feet long, 21½ to 56⅝ feet wide, and 1⅔ to 3⅙ feet draught loaded; and the barges are 81 to 223 feet long, 11½ to 30½ feet wide, and 3 to 9⅙ feet draught when loaded, and have a tonnage of 90 to 820 tons [1]. The aim, however, of the government in regulating the Danube and the Theiss, is to obtain a minimum depth at the lowest stage of 6½ feet, to allow of the navigation of vessels of 500 to 600 tons. The navigation of the rivers in Hungary is free of toll.

The direction of the flow of the Danube, and the position of its main tributaries, are not favourable for the river traffic, as the main traffic on the Danube, being from east to west, is up-steam, and the course of its tributaries necessitates circuitous routes to Buda-Pesth. Wheat, however, which constitutes the principal trade, is very conveniently carried in bulk by water; the traffic on the Danube constitutes about 80 per cent. of the whole, as the other waterways act as its feeders; and the grain trade nearly equals in weight the rest of the goods carried by water [2]. Fuel, building materials, and timber, compose about 20 per cent. of the traffic; whilst manufactured goods for the east are attracted by the low rates, and the easy rapid transit down-stream on the Danube.

**Traffic on Waterways and Railways in Hungary compared.**
On the introduction of railways into Hungary, attention was for many years exclusively directed to their development, the length of the railways having been increased from 138 miles in 1850, to 7,366 miles in 1892. This extension of railways has naturally diverted a large proportion of the passenger traffic from the rivers, except for short journeys; and it has greatly augmented the trade of the country. The fall, however, in

---

[1] 'Les Voies Navigables de la Hongrie,' Buda-Pesth, 1892, p. 12.

[2] 'Rôles respectifs des Voies Navigables et des Chemins de Fer dans l'Industrie des Transports en Hongrie,' A. Halasz, Vme Congrès International de Navigation intérieure, Paris, 1892; and 'The Utilisation of Water and Rail Routes in Hungary,' A. Halasz, Water Commerce Congress, Chicago, 1893.

the price of wheat since 1880, has led to a reduction in the railway rates, to enable Hungary to maintain its competition with foreign markets, and has directed the attention of the Government to the improvement of the waterways for facilitating cheap transport. The railway traffic measured by ton-miles is about three times the traffic on the rivers, which was 613 million ton-miles in 1890; whilst the actual tonnage of goods carried on the railways was about seven times that conveyed on the waterways. The increase in the total tonnage of goods between 1881 and 1890, was 53 per cent. on the railways, and 52 per cent. on the waterways; but the ton-mileage on the railways was increasing at a more rapid rate than on the waterways. The removal, however, of the obstacles to navigation on the Danube at various places in Hungary, and more especially at the 'Iron Gates,' just beyond the frontier, and the general improvement in depth which the regulation works in progress will secure, especially if supplemented by the construction of some important connecting links for shortening the routes, will give an impulse to the river traffic and improve the conditions of trade in Hungary.

## INLAND NAVIGATION IN RUSSIA.

The vast area, large rivers, extensive plains, and agricultural character of Russia, render it naturally suitable for inland navigation. The great lengths, however, of the rivers, the obstacles met with in some cases, the barrenness of certain districts, the scarcity of population in some parts, and the attention bestowed for many years to the development of railways, have caused the improvement of the waterways for navigation to be very slowly carried out.

**Russian Navigable Rivers.** The rivers of Russia flow into four quite separate seas, namely, the Arctic Ocean with the White Sea to the north, the Baltic Sea to the west, and the Black Sea and Caspian Sea to the south. The blocking of the northern seas by ice during nearly half the year hampers the trade on the rivers flowing into them, of which the principal are the Onega, the Dwina, and the Petchora in Russia in Europe, and the Obi, Yenisei, and Lena in Siberia. The Baltic is open for about three-fifths of the year; and the principal Russian rivers falling into it are the Neva, Duna, Niemen, and Vistula. The chief Russian rivers of the Black Sea and Sea of Azof are the Don, the Dnieper, and the Dniester, and of the Caspian, the Volga and the Ural. These rivers possess, for the most part, the important advantages of traversing a large extent of country with a moderate fall and a fairly regular discharge, and freedom from floods except on the breaking up of the ice in the spring. They are, moreover, only separated from one another by a comparatively short water-parting of moderate elevation, and are therefore easily connected by a canal; and they are more or less navigable in their natural condition for long distances inland. The Don, indeed, is impeded in the middle of its course by stony shoals; the Dnieper is obstructed by various rapids which require channels cut through the rocks, and locks in some places, to render the river convenient for navigation[1]; and the Dniester, in the absence of tributaries, has an inadequate depth, except during its somewhat frequent floods. The Volga, however, the largest river of Russia, is quite free from rapids for nearly the last 2,000 miles of its course; whilst the Dnieper is navigable for about 950 miles above its rapids, and for 220 miles below them. The Volga, Vistula, Duna, Dnieper, Dniester, and other rivers have been regulated from time to time at the worst places, with the object of increasing the navigable depth and preventing the falling in of the banks; but recently complete surveys and sections of the chief rivers have been made, and their rate of flow, discharge, and depth have been measured, with a view of

[1] 'L'Aperçu général des Voies Navigables de la Russie,' N. de Sytenko, Manchester Inland Navigation Congress, 1890; and 'Aperçu historique du Développement des Voies Navigables de l'Empire de Russie,' E. F. de Hoerschelmann, Kief, 1894.

instituting systematic works of improvement of the rivers for navigation, instead of dealing with isolated shoals. Some of the minor rivers have been rendered navigable by canalization. **Inland Canals in Russia.** Canals have been formed in Russia solely with the object of connecting together the chief navigable rivers, and thus forming unbroken waterways across the country from sea to sea; and their construction has been rendered easy by the peculiarly favourable conditions previously referred to. Thus the Caspian is connected with the Baltic by the Marie system of canals, uniting the Volga with the Neva, and thereby enabling the products of southern Russia to be conveyed by water to St. Petersburg, and by the Baltic to foreign ports. Two other shorter routes, indeed, have been formed for connecting the Caspian with the Baltic, by means of the short Tikhvinski Canal joining two canalized tributaries of the Volga and Lake Ladoga, and by the Vychni-Volotchok Canal, 30¾ miles long, joining the Tsna, a tributary of the Volga, with another river flowing into Lake Ladoga, which formed the only connection between the Volga and St. Petersburg up to the end of the eighteenth century. The incompleteness, however, of the works of the first route, and the rapids on one of the tributaries in the second route, have led to the diversion of the through traffic to the improved Marie system and to the railways.

The Marie navigation, which was originally formed by works commenced in 1799, starts from the Volga at Rybinsk, and follows the river Cheksna to Lake Bielo-Ozero, which it skirts by the Bielozerski Canal, and enters the river Kovja, which it joins to the Vytegra, a tributary of Lake Onega, by the Marie Canal crossing the water-parting of the basins of the Caspian and the Baltic [1]. The navigation then skirts Lake Onega by the Onejski Canal, this lake being connected with Lake Ladoga by the river Svir, and passing along this

---

[1] 'La Réorganisation du Système fluvial Marie,' E. F. de Hoerschelmann, Manchester Inland Navigation Congress, 1890.

river, and round the southern shore of Lake Ladoga by the
Ladojski Canals, it joins the Neva, and thus connects the
Baltic with the Caspian. The total length of this waterway,
from the Caspian Sea to St. Petersburg is 2,507 miles, and
of this length, 668 miles belong to the Marie Navigation, of
which about 200 miles consist of canals. This navigation has
been quite transformed by works recently carried out, consisting in the rebuilding of the locks, with an available length
of 1,050 feet and a width of 42 feet ; the canalization of the
Cheksna, where impeded by rapids, and its regulation at sharp
bends ; and the removal of rocks and dredging shoals, so as
to provide a minimum depth of 6½ feet. The summit-level
of the navigation on the Marie Canal, reconstructed at a
lower level in 1882–6, is now 420 feet above sea-level; and
there are only thirty-seven locks on the whole route, with
falls not exceeding 10½ feet. The navigation is now accessible
to vessels of 655 tons, 210 feet long, 31½ feet wide, and drawing
5⅝ feet of water.

The Baltic is connected with the Black Sea by the junction
of the Duna to the Dnieper, through their tributaries the
Oulla and the Beresina, by the Berizinski Canal across the
water-parting of their basins, constructed early in the nineteenth century. This waterway is 408 miles long, of which only
105 miles are canalized. Another connection of these seas has
been effected by the Oguinski Canal, about 30 miles long,
joining two tributaries of the Niemen and the Dnieper, by
traversing the water-parting of their basins. The canalization
has been effected by eleven locks and some needle weirs ;
and the chief traffic consists of rafts of timber conveyed to
Germany. A third route has been provided between the Baltic
and the Black Sea by the Dnieper-Bug Canal uniting these
rivers, and consequently the Dnieper and the Vistula, the
canalization being effected by twenty-two needle weirs ; and
the Vistula is connected directly with the Niemen, through
its tributary the Nareff, by the Augustof Canal.

The Duke Alexander of Wurtemburg Canal connects the river Cheksna with the river Soukhona through Lake Koubennskoïe, and consequently the Marie Navigation with the Dwina, and the Baltic and Caspian with the White Sea. The Catherine, Vindafski, and Volga-Moskva canals, uniting the Volga with the Dwina, the Niemen with the Vindau, and the Volga with the Moskva, through some tributaries, have been abandoned, owing to the large cost of their reconstruction in comparison with the prospects of traffic. As the Volga flows near the boundary of the Don basin between Saratoff and Tsarizine, the construction of a canal has been proposed, only 35 miles in length, for connecting the Volga with a tributary of the Don, and thereby to give an outlet from the Volga into the Black Sea for vessels of 500 tons.

Most of the Russian canals and river improvement works in connection with them, were carried out during the latter half of the eighteenth century and the first half of the nineteenth century; and whilst some of these navigations have been enlarged and the locks rebuilt, others remain in their primitive condition with wooden locks. Since the development of the railway system in Russia, the only important works of canalization which have been carried out are the enlargement of the Marie Navigation, and the canalization of the river Moskva from Moscow to its confluence with the Oka, a tributary of the Volga, in Russia, and the Obi-Yenisei Canal in Siberia, connecting tributaries of those rivers.

**Lengths of Russian Waterways.** The total length of navigable waterways in Russia in Europe amounts to about 53,000 miles, of which 43,000 miles are under the control of the State; but it appears that between one-fourth and one-fifth of the whole length is only available for rafts of timber floated down [1]. Assuming that about 12,000 miles

[1] 'L'Aperçu général des Voies Navigables de la Russie,' N. de Sytenko, Manchester Inland Navigation Congress, 1890, p. 6.

are only floatable, there would remain 41,000 miles of waterways more or less navigable; and this, with an area of 2,095,500 square miles for Russia in Europe, would give nearly 2 miles per 100 square miles, which though a smaller proportion than that of the French or German navigable waterways, is large considering the vastness of the country and the small length of the canals. The length of the navigable rivers in Siberia has not been ascertained; but surveys of some of the rivers are being made, and some regulation works are being carried out.

**Traffic on Russian Waterways.** The actual traffic on Russian waterways is not exactly known, but it has been estimated at about 32,250,000 tons, or rather a less weight of goods than carried by water in England[1]. The average distance traversed, however, estimated at 1,000 versts or 663 miles, is naturally much larger on the very long waterways of Russia than in any other country of Europe; for even in Germany, with its large rivers, the average journey is 217 miles; whilst in France it is 88 miles, and in England it has been estimated at only 35 miles. Accordingly, the yearly traffic in Russia reaches about 21,400,000,000 ton-miles, nearly ten times the ton-mileage on the French waterways, and about seven times the ton-mileage on the German waterways.

### INLAND NAVIGATION IN INDIA.

The plains of India which are at a sufficiently·low level to be suitable for navigation, are confined to the basins of the Indus, the Ganges, and the Brahmaputra, the strip of land bordering the east coast from the delta of the Ganges down to Cape Comorin, and the valley of the Irrawaddi in Burmah.

**Navigable Rivers in India.** Inland navigation on the rivers of India is almost wholly confined to the Ganges, the Brahmaputra, the Indus, and the Irrawaddi, with some of

---

[1] 'Aperçu historique du Développement des Voies Navigables de l'Empire de Russie,' E. F. de Hoerschelmann, Kief, 1894, p. 58.

their tributaries, which have served from time immemorial for the transport of the produce of the interior to the sea-coast, or to the towns on the banks of these rivers. The Ganges becomes navigable on emerging from the hills at Hudwar; the Brahmaputra is navigable for steamers below Dibrugarh, 800 miles from the sea; the Indus commences to be navigable near Attock, 940 miles from its mouth; and the Irrawaddi is navigated by small boats for a distance of 700 miles. The opening, however, of the East Indian Railway diverted the traffic from the steamers on the Ganges; whilst the Indus Valley Railway greatly reduced the steam traffic on the Indus. River navigation, nevertheless, continues to flourish extensively on the numerous channels of the delta of the Ganges; and the Brahmaputra and Irrawaddi retain their importance as great highways of trade.

**Navigable Canals in India.** Several irrigation canals in India have been constructed so as to be available for navigation along a portion of their course. Thus in the Punjab, the Western Jumna and Sirhind canals are navigable for 432 miles; in the North-west Provinces, the Agra, Ganges, and Lower Ganges canals are navigable for 512 miles; in Bengal, the Sone, Midnapore, and Orissa canals are navigable for 467 miles; and in Madras, the Godavery, Kistna, and Kurnool canals are navigable for 970 miles[1]. The canals on which navigation has been combined with irrigation with the most successful results, are the Sone, Midnapore, Orissa, and Godavery canals, on which trade is carried on regularly by steamers. These canals, with the exception of the Sone Canals, are deltaic canals; and the Sone Canals also traverse very flat country.

The only canals in India which have been constructed specially for navigation are the Calcutta and Eastern Canal, the Orissa Coast Canal, and the Buckingham Canal in Madras. The Calcutta and Eastern Canal starts from the

---

[1] 'Irrigation Works in India and Egypt,' R. B. Buckley, p. 212.

Hooghly at Calcutta, and goes eastwards across the delta of the Ganges, connecting Calcutta with the eastern districts, and also with the northern districts by the delta channels which the canal traverses, thereby enabling the products of these districts to be brought by water to Calcutta. This canal was partially opened at the close of the eighteenth century, and returns a profit of nearly 4 per cent. on its capital cost. The Orissa Coast Canal, running fairly parallel to the coast, connects Orissa with the Hooghly, and consequently with Calcutta. It was completed in 1887; but its revenue does not suffice to pay its working expenses. The Buckingham Canal starts from the Kistna delta, and following very closely along the coast-line, crosses the Penner delta, and passing Madras proceeds southwards, terminating a little south of the river Palar, about twenty-five miles north of Pondicherry. This canal was undertaken after the famine of 1877–78, as a protective work; and its revenue does not quite cover the expenses of working.

## INLAND NAVIGATION IN CANADA.

The river St. Lawrence, and the remarkable chain of large inland lakes which it drains, furnish Canada with the finest inland navigation system in the world; though canals have had to be constructed at some parts to surmount the obstacles presented by rocky rapids in the rivers connecting the lakes, and also along portions of the St. Lawrence itself above Montreal.

**Waterways of Canada.** The St. Lawrence and the lakes Ontario, Erie, St. Clair, Huron, and Superior, with their connections, provide a large inland waterway penetrating 2,260 miles into the interior up to Port Arthur at the head of Lake Superior, with a rise of about 600 feet; and the same waterway furnishes Chicago direct access to the Atlantic through Lake Michigan. Six lateral canals have been formed alongside the St. Lawrence, above Montreal, for improving the

navigation. The Lachine Canal, 8½ miles long and having
five locks, cuts across a great bend of the river, just above
Montreal, thereby avoiding the St. Louis rapids, the first
obstacle to the ascent of the St. Lawrence, and shortening the
route. The Beauharnois Canal, 11¼ miles long and provided
with nine locks, is the only lateral canal on the right bank of
the river, and connecting lakes St. Louis and St. Francis,
enables vessels to avoid three rapids on the river. The
other lateral canals are the Cornwall, Farran's Point, Rapide
Plat, and Galops canals, between Cornwall and Prescott, 11½,
¾, 4, and 7⅝ miles long respectively, and having altogether
twelve locks to allow for the rise in the river [1].

The Welland Canal was constructed to connect Lake Ontario
at Port Dalhousie with Lake Erie at Port Colborne, as the
Niagara River is barred by its rapids and falls; and the canal
is 26¾ miles long, and has twenty-six locks to surmount the
difference in level of 326¾ feet. The Sault-Sainte-Marie Canal
is being constructed through St. Mary's Island, to avoid the
rapids of the river St. Mary, by a cut 3,500 feet long, with
one lock, which will provide a navigable passage between
Lake Huron and Lake Superior on the Canadian side; whereas
hitherto vessels have had to pass through the St. Mary's Falls
Canal on the United States bank of the river (p. 376).

The river Richelieu, flowing into the St. Lawrence at Sorel,
46 miles below Montreal, connects the St. Lawrence with
Lake Champlain, and consequently with New York by the
Champlain Canal and the Hudson River, the distance between
Sorel and New York by this route being 411 miles, of which
81 miles are in Canadian territory. The navigation of the
river Richelieu has been improved by a lock at St. Ours in
a cut 1 furlong in length, and by the lateral Chambly Canal,
12 miles long, with nine locks rising 74 feet, to avoid the
rapids between Chambly and St. John's.

[1] 'Dominion of Canada, Annual Report of the Department of Railways and
Canals for the year 1893-94,' Ottawa, 1895, pp. lxiv to lxxxvii, and maps.

Ottawa is in communication by water with Montreal by the river Ottawa flowing into the St. Lawrence above Lachine, and with Kingston at the eastern end of Lake Ontario, by the Rideau navigation. The navigation of the river Ottawa has been improved by a lock at St. Anne's in a channel a furlong in length, the Carillon Canal three-quarters of a mile long, and the Grenville Canal 5¾ miles long with five locks; and the length of the route from Montreal to Ottawa is 119⅝ miles. The Rideau navigation, 126¼ miles long, comprises the river Rideau, the Rideau Canal, and the river Cataraqui; it rises 282 feet from Ottawa by thirty-five locks, and descends 164 feet to Kingston by fourteen locks; and its summit-level is Lake Rideau.

The Murray Canal is an open cut without locks, 5⅛ miles long, across the isthmus of Murray, connecting the head of the Bay of Quinté directly with Lake Ontario, and thereby enabling vessels to avoid the open lake. The Trent navigation comprises a chain of lakes and rivers, extending from Trenton at the outlet of the river Trent into the Bay of Quinté near the Murray Canal, to Lake Simcoe, and thence by the Severn River to Georgian Bay opening into Lake Huron. It was proposed many years ago to form a navigable connection between Lake Ontario and Lake Huron by this route, about 235 miles in length; but at present this system is merely of local value, sections only of the route having been rendered available for navigation by lateral canals, locks, and dams, for avoiding the rapids. The longest portion of continuous navigation is from Lakefield to Balsam Lake, 61 miles in length, which is being extended about 4 miles at each extremity; whilst further extensions have been surveyed.

**Dimensions of Canadian Canals and Locks.** The available capacities of the waterways of Canada are determined by the sizes of the artificial links which overcome the natural obstacles presented by the rivers; and the increase in trade, and in the sizes of vessels traversing the principal through routes, have necessitated successive enlargements of the main connecting links. The lateral canals, with their locks, on the route between the lakes and Montreal, are being gradually enlarged or reconstructed, so as to provide a navigable depth of 14 feet in place of 9 feet, and locks 270 feet long and 45 feet wide. The enlargement of the Welland Canal has been completed, and also of the Lachine Canal, which has been given a mean width of 150 feet; and the Soulanges Canal, 14 miles long, with five locks of the standard dimensions, is being constructed along the north bank of the St. Lawrence above the confluence of the Ottawa, to replace the Beauharnois Canal on the opposite bank,

The Sault-Sainte-Marie Canal, which forms a link in the lake system, completed in 1894, has had its dimensions modified during construction, so as to accommodate vessels drawing 20 feet at the lowest water-level; its depth is 22 feet, and its bottom width is 145 feet; whilst its lock is 900 feet long and 60 feet wide, and affords a minimum depth of water of 20¼ feet over the sills. The lock will thus admit one lake vessel, 320 feet long, and two of the vessels navigating the Welland Canal, 255 feet long, at the same time.

The canals of the river Ottawa have locks 200 feet long and 45 feet wide, with a depth of 9 feet of water over the sills. The St. Ours lock on the Richelieu navigation has also a length of 200 feet and a width of 45 feet, but the available depth is only 7 feet; whilst the smallest lock on the Chambly Canal, forming a section of this route, is only 118 feet by 22½ feet. The locks along the Rideau navigation are 134 feet by 33 feet, and 5 feet depth on the sills; but the waterway has an available depth of only 4½ feet.

**Traffic on Canadian Waterways.** The largest quantity of traffic passes through the Welland Canal, amounting to 1,294,823 tons in 1893, of which 281,583 tons were carried up and 1,013,240 tons down[1]. On the St. Lawrence Canals, the quantities were 272,536 tons up and 885,840 tons down, making a total of 1,158,376 in 1893; on the Chambly Canal 192,324 tons went up and 120,546 tons down, or a total of 312,870 tons; on the Ottawa Canals 1,049 tons passed up and 580,472 tons down, giving a total of 581,521 tons; and on the Rideau Canal, the traffic was 79,653 tons up and 24,581 tons down, with a total of 104,234 tons in 1893.

The traffic on the Welland Canal has remained fairly uniform with occasional fluctuations, and was higher in 1893 than in any previous year since 1874, when the traffic was 1,389,173 tons. The through traffic from Lake Erie to Montreal, along the Welland Canal and the St. Lawrence, has risen from 169,000 tons in 1881 to 508,000 tons in 1893. Though the traffic on the several routes does not appear large in comparison with the returns of some of the European waterways, it must be remembered that the goods are carried for long distances by water in Canada, as in Russia, and that the local lake traffic, and the local traffic on the rivers which does not pass through the canals are not included in the returns.

Owing to the climate of Canada, its waterways have the disadvantage of being closed by ice for rather more than four months in the winter; but, nevertheless, the great natural facilities for navigation afforded by the rivers, and the extensive tracts of fertile lands served by the lakes, enable the waterways, with their enlarged capacities, to maintain a profitable competition with the railways.

### INLAND NAVIGATION IN THE UNITED STATES.

The United States shares with Canada the possession of four of the great lakes, whilst Lake Michigan is wholly within its territory; and though the St. Lawrence below Cornwall passes into Canada, the United States possesses in the Mississippi

---

[1] 'Dominion of Canada, Annual Report of the Department of Railways and Canals for the year 1893-94,' pp. 254, 255, and 392.

one of the largest navigable rivers of the world. The United
States, indeed, with its vast extent of territory, with large
districts rich in agricultural produce and other resources at
a long distance from the sea-coast, with its large chain of
inland lakes, and with several large rivers navigable for long
distances inland, is a country where inland navigation might
be expected to attain its highest development.

**Navigable Rivers of the United States.** The Mississippi is
navigable from its mouth up to the St. Anthony Falls near
St. Paul, a distance of about 1,850 miles. It has had its
minimum depth gradually increased by training works and
dredging shoals; and now, below the confluence of the Red
River, its navigable depth is never less than 10 feet; but
sometimes in the first 550 miles below Cairo, it is reduced to
5 or 6 feet over the shoals. The Upper Mississippi is also
being improved by training works and the removal of
shoals; whilst the erection of locks and weirs between
Minneapolis and St. Paul is being undertaken to maintain
a better depth [1]. The Des Moines rapids, above the confluence
of the Des Moines River, have been avoided by a lateral
canal on the right bank, 7⅔ miles long, 250 to 300 feet wide,
and 5 feet minimum depth, with three locks 334 feet long
and 80 feet wide. The Rock Island rapids, a short distance
below the outlet of the Hennepin Canal, causing a fall of
21½ feet in 14 miles, have been improved by cutting channels
through the seven limestone reefs, the total length of the cuts
amounting to 3 miles [2]. The other navigable rivers of the
Mississippi basin comprise the Ohio up to Pittsburg, 967 miles
from its confluence with the Mississippi at Cairo, and portions
of its tributaries, the Tennessee, Cumberland, Green, Wabash,
Kentucky, Great Kanawha, Monongahela, and Alleghany
rivers; the Missouri River, from its confluence with the

---

[1] 'Report of the Chief of Engineers, U.S.A., for 1894,' part 3, pp. 1640
and 1681.

[2] Minutes of Proceedings, Institution C. E., vol. xl. p. 188, and plate 6.

Mississippi above St. Louis to the mouth of the Marias River half way across Montana, which is notable rather for the length of its navigation than for its traffic; the Illinois River which provides an outlet for the traffic on the Illinois and Michigan Canal; and the Red River, and other minor tributaries of the Mississippi[1].

The Hudson River connects New York with the Erie and Champlain canals, and consequently with the lakes and the St. Lawrence. Some minor rivers flowing into the Gulf of Mexico to the east of the Mississippi delta, and the St. John's, Altamaha, Savannah, and numerous other rivers flowing into the Atlantic, are navigable for about 30 to 200 miles inland; whilst the James River is navigable up to Richmond, and the Delaware River up to Philadelphia.

Most of the above rivers have been improved by training works, protection of banks, and dredging; whilst the Falls of the Ohio at Louisville have been avoided by the lateral Louisville and Portland Canal, about 2 miles in length, with a flight of two locks having chambers 372 feet long and 80 feet wide, and a total lift of 26 feet. Several of the rivers have had their navigable capabilities extended by locks and weirs, as for instance the Ohio, Cumberland, Green, Kentucky, Great Kanawha, Monongahela, Alleghany, and Illinois rivers.

Inland Navigable Canals in the United States. The large number of rivers navigable for long distances in North America, rendered the construction of canals as means of communication before the introduction of railways unnecessary, except for the purpose of forming connecting links in long lines of waterways. Thus the canals constructed in the United States in the first half of the nineteenth century were the Erie Canal, extending from Troy on the Hudson River along the Mohawk valley, thereby avoiding the Alleghanies, to Buffalo on Lake Erie, 352 miles long, with a branch to Oswego on Lake Ontario,

---

[1] 'Outline Map of the United States and Territories, showing the Tonnage of the Navigable Rivers, 1890,' supplied to me by the Chief of Engineers.

38 miles long, and thereby connecting New York by water with the lakes; the Champlain Canal from Albany to Lake Champlain, 66 miles long, which by way of the river Richelieu provides a direct waterway between New York and the St. Lawrence; the Morris Canal joining the Hudson at Jersey City opposite New York with the Delaware at Philipsburg, 102 miles long; and the Illinois and Michigan Canal from Chicago on Lake Michigan to La Salle on the Illinois River, 97 miles long, connecting the lakes with the Mississippi, and consequently with the Gulf of Mexico, which completes a waterway extending down the whole length of the United States[1]. Another canal partially constructed during the same period, was the Chesapeake and Ohio Canal, designed for connecting the Potomac at Georgetown, a suburb of Washington, with the Ohio at Pittsburg, having a total length of 341 miles, so as to form a continuous waterway extending from the Atlantic at Chesapeake Bay to the Mississippi, and thereby communicating with the lakes to the north and the Gulf of Mexico to the south. This canal, however, was only completed as far as Cumberland in the Potomac valley by 1850, for a length of 186 miles; and though proposals have been made for crossing the ridge of the Alleghanies separating the valleys of the Potomac and the Yougiogheny, a tributary of the Ohio, by a series of inclines rising 1,185 feet from Cumberland, and traversing the top of the ridge by a tunnel at the summit-level, $3\frac{3}{4}$ miles long, the canal has not been extended. A second canal, to the north of the Chesapeake and Ohio Canal, was constructed in a similar direction, from Havre de Grace at the northern extremity of Chesapeake Bay, along the valleys of the Susquehanna and Juniata rivers to Frankstown, which it was proposed to extend to Johnstown about the same distance from Pittsburg as Cumberland; but this canal appears to have been

---

[1] 'Map showing the Location of Works and Surveys for River and Harbour Improvement, 1879,' supplied to me by the Chief of Engineers.

sold to the Pennsylvania Railway Company, and practically abandoned.

Two main waterways were recommended, by a Select Committee in 1874, to be formed for connecting the Mississippi valley directly with the Atlantic, known respectively as the Central and Southern Transportation Routes[1]. The Central Transportation Route starts from the Ohio at Point Pleasant, along the Great Kanawha River, which is being canalized by locks and movable dams (p. 139); and following the New and Greenbrier rivers, it is designed to pierce the Alleghany ridge, separating the basins of the Mississippi and James rivers, in a tunnel nearly 8 miles long; and then by a canal along the James valley, it would connect the western end of the James River with the Kanawha Canal, which has been constructed between Buchanan and Richmond, a distance of 197½ miles. The total length of the route between the Ohio and Richmond is 471½ miles, of which 231 miles would be canal, 161½ miles canalized river, and 79 miles of open river. The rise from the Great Kanawha River to the proposed summit-level is 1,114 feet, and the fall from the summit-level to Richmond is 1,700 feet; and Richmond is in direct communication with the Atlantic by the James River. The traffic on the Great Kanawha River, especially in coals, has greatly increased with the progress of the improvement works; but it appears doubtful whether the traffic on the through route, in the face of railway competition, would at all compensate for the great expenditure involved in the tunnel and other works for completing the waterway.

The Southern Transportation Route is designed to connect the Tennessee River at Guntersville with Macon on the Ocmulgee River, a tributary of the Altamaha which flows into the Atlantic near Darien. The proposed route passing by canal and canalized river to Gadsden, then to Rome by

[1] 'Report of the Select Committee on Transportation Routes to the Seaboard,' Washington, 1874.

the Cossa River and on to Macon by canalized river and canal, has a total length of 415⅗ miles, of which 70½ miles are canalized river, 192¼ miles canal, and 153 miles open river. This proposed route, which would rise 400 feet and fall 464 feet in crossing from the valley of the Tennessee River to that of the Cossa River, and ascend 266 feet and descend 705 feet in passing from the Cossa valley to the Ocmulgee valley, has been commenced by the improvement of the Cossa River between Gadsden and Rome, a distance of 153 miles, constituting the simplest portion of the route [1]. The approaches, however, of this route on each side have been improved by the construction, in 1875–90, of the Muscle Shoals Canal in two divisions, for avoiding the Muscle and Elk shoals on the Tennessee River, and the improvement of the Ocmulgee and Altamaha rivers. The upper portion of the Muscle Shoals Canal is 1½ miles long, and has two locks 300 feet long and 60 feet wide, with a total rise of 23 feet; and the lower division is 14½ miles long, with nine locks of similar size, and a total rise of 85 feet [2].

Two short canals, or cuts, have been formed for improving the navigable waterway between Lake Superior and Lake Huron, at Sault-Sainte-Marie, and at the outlet of the St. Clair River into Lake St. Clair. The first canal, known now as the St. Mary's Falls Canal, avoiding the falls on St. Mary's River, has been already described (p. 376); and its successive enlargements have been necessitated by its forming the sole outlet for the traffic on Lake Superior. The St. Clair Flats Canal was constructed in 1866–71, to provide a direct and deeper outlet into Lake St. Clair than any of the seven mouths of the river afforded. The channel was first made 13 feet in depth, it was deepened in 1873 by dredging to 16 feet, and is now being deepened to 18 feet, which eventually is to be carried

---

[1] 'Report of the Chief of Engineers, U.S.A., for 1894,' part 2, p. 1286.
[2] Ibid. part 3, p. 1828.

to 20 feet, thereby improving the communication between Lake Huron and Lake Erie.

A canal is in course of construction for connecting the Upper Illinois River near Hennepin with the Mississippi River at the confluence of the Rock River, in order to provide a deeper and more direct waterway between Lake Michigan and the Mississippi than the Illinois River, in connection with the enlarged waterway which is being formed from Chicago along the line of the Illinois and Michigan Canal [1]. This Illinois and Mississippi, or Hennepin Canal, 77 miles long, is designed to rise 205 feet by twenty-four locks from the Illinois River to the summit-level, and to descend 102 feet by thirteen locks to the Mississippi. The canal is being formed 80 feet wide at the water-level, and 7 feet deep ; and the locks are to be 170 feet long and 35 feet wide. The length of actual canal is 50 miles, the remaining 27 miles being formed by canalizing the Rock River; and the works were commenced in 1892 at the mouth of this river.

Several cuts have been made for connecting some of the numerous creeks, estuaries, and lakes near the Atlantic coast, between the mouth of James River and Pamlico Sound, such as the Dismal Swamp and Chesapeake and Albany canals; but these channels serve rather to facilitate the coasting trade than inland navigation, of which the Harlem Ship-Canal, opened in 1895, is an example on a large scale, connecting the Hudson River with Long Island Sound through the Harlem River.

**Chicago Drainage Canal.** Chicago has for many years discharged its sewage into Lake Michigan, whose waters have at the same time served as its source of water-supply. Though this joint system has been rendered less obnoxious by drawing the supply of water through tunnels under the lake with their inlets at some distance from the shore, the great increase in the population of Chicago, owing to its

---

[1] 'Report of the Chief of Engineers, U. S. A., for 1894, part 4, p. 2162.

unrivalled position at the head of the great chain of lakes, has imposed on the municipality the necessity of obtaining another outlet for its sewage. A large drainage canal is, accordingly, being constructed for conveying the sewage of Chicago away from the lake and the Chicago River, into the Illinois River, a course which has been rendered practicable by the Des Plaines River, a tributary of the Illinois River, which the canal joins near Lockport about 35 miles from the lake, having its level several feet lower than the water leve of the lake ; so that by excavating through the low intervening ridge, it has been possible to give the canal a fall of from 1 in 40,000 to 1 in 20,000 to its outlet into the Des Plaines River[1]. The river is to be improved to Joliet, making the total length of the work 38 miles ; and from this point the valley falls rapidly for some distance towards the Illinois River. The canal has been designed to be suitable for navigation for the vessels traversing the lakes, as well as for its primary object of drainage, in order to overcome the opposition of the inhabitants of the Illinois valley, by providing them with a waterway to the lakes, which it is hoped will be eventually extended down the Illinois River so as to provide an ample waterway between the lakes and the Gulf of Mexico. Accordingly, the canal has been given a bottom width of 160 feet in rock, and 202 feet in clay, with side slopes of 2 to 1 ; and sufficient water will be drawn from Lake Michigan to provide a depth of 22 feet, which may eventually be increased to 26 feet. The works were commenced in 1892, and are to be completed in 1896 ; and a special feature in the plant for carrying out the work, is the employment of balanced cantilever derricks, 342 feet long, which can each deposit 600 cubic yards of excavated material per day over

---

[1] 'The Proposed Enlargement of the Waterway from Lake Michigan to the Mississippi River via the Illinois River,' L. E. Cooley, Water Commerce Congress, Chicago, 1893; and 'Drainage Channel and Waterway,' G. P. Brown, Chicago, 1894, pp. 374 and 416, and plate.

a spoil bank at the side, which may be raised 90 feet high ; whilst excavators and steam navvies have been largely used for excavating the channel, together with high-power derricks and conveying bands for the removal of the materials. Some locks and weirs will be necessary to extend the navigation down to the Illinois River ; and the canalization of the Illinois River will have to be carried out to complete the waterway to the Mississippi.

**Sizes of Navigable Waterways in the United States.** Along parts of many of the large rivers of the United States the navigation is open, and the size of the vessels that can traverse them is only limited by the available depth, which is being gradually increased by training works, dredging, and the removal of obstructions. Where however, canalization is introduced for improving the navigation, the size adopted for the locks determines the capabilities of the waterway. Thus in the Great Kanawha River, the locks are 300 feet long and 50 feet wide above Charleston, and 340 feet long and 55 feet wide below ; whilst the locks on the Cossa River are only about 200 feet long and 40 feet wide ; and the locks about to be constructed on the Cumberland River are designed to be 280 feet long and 52 feet wide. Still larger locks, however, are proposed for the canalization of the Upper Mississippi between Minneapolis and St. Paul, with a length of 334 feet, a width of 80 feet, and a minimum depth at low water of 5 feet.

The early canals in the United States were made of small section, as in other countries. The Erie Canal was originally formed with a bottom width of 28 feet, a width at the water-level of 40 feet, and a depth of 4 feet, and was accessible to vessels of 60 tons; but it was subsequently enlarged to 56 and 70 feet, with a depth of 7 feet, and locks 110 feet long and 18 feet wide, admitting vessels of 240 tons. Proposals also have been made for forming an enlarged waterway between the Hudson River and Lake Erie, more suited to the increased demands of traffic of the present day, and capable of competing more successfully with the railways for the carriage of grain and other bulky goods from the lakes to New York. The Morris Canal was originally constructed only 20 feet wide at the bottom and 4 feet deep, so that it had to be enlarged in 1841–45 to 25 feet bottom width and 5 feet depth, to be capable of admitting vessels of 44 tons; and it was enlarged again in 1860, so as to be navigable by vessels of 80 tons. Its chief interest, however, now consists in its inclines (p. 392), for its leasing to the Lehigh Valley Railway Company in 1870 deprived it of any importance as a navigable waterway[1]. The Chesapeake and Ohio Canal has the same dimensions as the enlarged Erie Canal; whilst the Illinois and Michigan Canal has a bottom width of 46 feet, a width of 60 feet at the water-level, and a depth of 7 feet; and the Champlain Canal had a bottom width of 35 feet and a depth of 5 feet, which have been increased to 44 feet and 7 feet. The further enlargement of this last canal has been proposed, with the object of providing a commodious waterway between New York and the St. Lawrence.

The lateral canals more recently constructed by the United States Government, to facilitate and extend the navigation along the large rivers, have been made of more ample dimensions, their lengths being small in proportion to the extent and importance of the navigations which they connect. Thus the Muscle Shoals Canal of the Tennessee River has locks 300 feet long and 60 feet wide; the Des Moines Rapids Canal of the Upper Mississippi has locks 334 feet long and 80 feet wide; the Louisville and Portland Canal of the Ohio river has locks 372 feet long and 80 feet wide; the Cascades Canal of the Columbia River is to be provided with a lock 462 feet long, 90 feet wide, and having a lift of 24 feet; and the new lock of the St. Mary's Falls Canal is being made 800 feet long, 100 feet wide, and 21 feet available depth.

---

[1] 'Notes on some of the chief Navigable Rivers and Canals in the United States and Canada,' G. T. Walch, Madras, 1877, p. 4.

The Hennepin Canal, indeed, which is also being carried out by the Government, with a bottom width of 59 feet and a depth of 7 feet, is little larger than the Erie Canal ; but if the Chicago Drainage Canal is extended by the canalization of the Illinois River, an ample waterway will be provided between the lakes and the Mississippi independent of this canal.

**Traffic on the Waterways of the United States.** The traffic on the Erie Canal, and the other canals belonging to the State of New York, increased fairly regularly up to about 1872 ; but since that period the traffic in vegetable food has somewhat diminished, whilst the traffic in iron, salt, coals, and ores, has been reduced to one half. Moreover, whereas in 1872 these canals carried slightly over 40 per cent. of the total tonnage carried by canals and railways in the State, the canal traffic was reduced to rather under 10 per cent; of the whole by 1892 ; for of the total traffic of 16,632,000 tons in 1872, as much as 6,726,000 tons were carried by the canals; whereas in 1892, out of a total traffic of 43,618,000 tons, only 4,283,000 tons were conveyed along the canals. This diminution in canal traffic, in spite of the large increase in trade, shows that the canals need enlarging, like the St. Lawrence Canals, in order to secure a fair share of the traffic, which is being diverted to the railways in the absence of improved facilities for navigation.

The rivers of the United States possessing the largest traffic are the St. Clair and Detroit rivers, on the route between Lake Huron and Lake Erie, and therefore forming links in the lake navigation, with 20,000,000 and 19,646,000 tons respectively in 1890 ; whilst the St. Mary River, forming the outlet for the traffic from Lake Superior, has a traffic of 8,288,000 tons ; and the Saginaw River, affording East Saginaw access to Lake Huron, has a traffic of 2,000,000 tons [1]. The Hudson River, between New York and the entrance to

[1] 'Outline Map of the United States and Territories, showing the Tonnage of the Navigable Rivers, 1890.'

the Erie Canal, holds the third place in the river traffic, with
18,582,000 tons in 1890, conveyed a considerably longer
distance than on the lake rivers. The navigable length, however, of the Hudson River is small compared to the distances
along which goods are conveyed on the Mississippi, the Ohio,
the Missouri, and the Arkansas. The traffic is largest on the
Ohio between Cairo, at its confluence with the Mississippi, and
Pittsburg, which amounted to 6,000,000 tons in 1890; and
the next most frequented portions of these rivers are the
Mississippi between the Illinois River and Cairo with 4,131,000
tons, between New Orleans and Cairo with 3,180,000 tons,
between St. Paul and the Illinois River with 3,000,000 tons,
and between New Orleans and the mouth of the Mississippi
with 2,862,000 tons. The Arkansas River possesses a traffic of
1,702,000 tons between Wichita in Kansas and its junction
with the Mississippi; but the Missouri, though navigable to
some extent for a longer distance than even the Mississippi, only
possesses a traffic of 849,000 tons between the Mississippi and
Sio at the mouth of the Big Sioux River, which is reduced to
14,600 tons higher up, and to 2,100 tons above Bismarck. Some
of the minor rivers of the Mississippi basin have a fair traffic,
though along shorter distances, namely the Monongahela
with 3,318,000 tons, the Great Kanawha with 1,127,000 tons,
the Alleghany with 973,000 tons, the Green River with 907,000
tons, and the Cumberland with 563,000 tons.

Some of the creeks, estuaries, and rivers flowing into the
Atlantic have a good traffic for comparatively short distances.
Thus the Harlem River at New York has a traffic of 3,384,000
tons, the Delaware to Philadelphia 4,193,000 tons, the Patapsco
to Baltimore 5,335,000 tons, the James River to Richmond
1,923,000 tons, the Savannah to Savannah 1,902,000 tons,
the Kennebec and Penobscot in Maine 2,549,000 and 2,165,000
tons respectively, the Charles River above Boston 1,500,000
tons, and the Connecticut River 1,070,000 tons. None of the
rivers of the United States flowing into the Gulf of Mexico,

except the Mississippi, have a large traffic, for the Escambia possesses the maximum traffic of only 345,000 tons.

The proximity of the Rocky Mountains to the Pacific prevents the rivers of the west coast having large basins or a long course. Nevertheless, the Sacramento River possesses a traffic of 1,006,000 tons from some distance above Sacramento to its mouth, and the Columbia River has a traffic of 1,633,000 tons up to Portland ; but the western river traffic is practically confined to these rivers and two or three of their tributaries.

Concluding Remarks. The eastern portion of the United States is well supplied with large rivers, which, together with the chain of lakes to the north, afford great scope for inland navigation ; and though the Alleghany range, stretching from north to south for a long distance, has hitherto prevented the direct connection of any of the rivers flowing into the Atlantic with the rivers of the Mississippi basin, the favourable conditions of the Mississippi and Ohio have attracted a large traffic over long distances inland ; and the works in progress for extending the navigation along the principal rivers and their chief tributaries, will promote the traffic. The rivers of the United States are fortunately under the sole control of the department of engineers of the United States army, ensuring uniformity and comprehensiveness in the improvement works and management, like in most European countries. The lengths, however, of the waterways are so great, like in Russia and Siberia, the needed works of improvement are so numerous and extensive, and the yearly appropriations granted by Congress are so often inadequate for the proposed works, that the progress seems slow compared with the urgency of the improvements ; and several schemes will require many years for their completion.

The connection of the Mississippi with Lake Michigan by an adequate waterway would be very valuable by uniting the two great arteries of inland navigation in the United States ;

and it appears capable of realisation in view of the completion of the most difficult portion by the construction of the Chicago Drainage Canal. The enlargement of the Erie Canal, which has been too long delayed, would confer a great benefit on the traffic by water between the great lakes and New York. The Central Transportation Route would afford a very useful connection by water between the Ohio and the Atlantic ; but the natural difficulties of the country to be traversed, culminating in a tunnel longer than the Mont Cenis, seem to preclude the completion of this proposed waterway with any prospect of advantages commensurate with the cost. The Southern Transportation Route appears to involve fewer difficulties in construction, and would provide an advantageous connection between the Mississippi and the Atlantic; but its completion would necessitate a very large outlay; its value would be less now, owing to the deepening of the outlet of the Mississippi, than when it was proposed in 1874 ; and at the present rate of progress, centuries would be required to finish it A more important object at the present day than the completion of these routes, appears to be to extend the navigation up the principal rivers, and to deepen the existing waterways. The Western States, owing to their unfavourable physical conditions for waterways, will always have to depend almost wholly on railways for inland communications.

# CHAPTER XXII.

## FORMS OF BARGES; METHODS OF TRACTION; AND SHIP-CANALS INSTEAD OF RIVERS.

THE forms and sizes of vessels, or barges, navigating rivers and canals, and the methods of traction employed, are important elements in the cost of working, the capabilities, the speed, and the convenience of inland navigation. They necessarily depend largely upon the size of the waterway, the nature of the traffic, the distance to be traversed, and any special conditions, such as a rapid current, or the necessity of traversing an exposed channel.

## FORMS OF BARGES.

The size of the barges intended to navigate any special waterway is limited by the minimum available depth of the waterway, and the width and length of the smallest lock ; and the barges must not approximate very closely to the available depth and width, in order to avoid an undue resistance to traction. Economical considerations would dictate that the barges should be given the maximum capacity consistent with the above limitations, which would be effected by giving them the same width and depth throughout. As, however, this rectangular form would evidently produce too great a resistance to traction, barges, though made with parallel sides and flat-bottomed, are slightly curved at the bow and stern. Where the depth is very moderate, as in large open regulated rivers, the deficiency in draught is made up by an increase in breath and length ; and sharp bends in such rivers are eased, in order to provide for the passage of long barges   This is the system which has been adopted on the Rhine, the Main, and other large rivers ; and where, in rare instances, locks are introduced on these navigations, they are made specially long.

The resistance to the motion of vessels in an ample waterway was shown by Mr. William Froude to be composed of surface friction, eddy resistance, and wave resistance due to the waves raised by the vessel[1]. Surface friction is proportionate to the surface in contact with the water, and varies with the roughness of the surface; eddy resistance may be practically eliminated by designing a ship with fine lines from the centre towards the bow and stern ; and wave resistance depends on the speed and length of the vessel.

[1] 'British Association Report, Bristol Meeting, 1875,' Transactions of Sections, p. 233.

Barges on rivers and canals are not intended to travel at high speeds, so that the wave resistance is small, except in restricted channels; and the length of the barges, and their rate of travel cannot be conveniently modified. As smooth a surface as practicable is advantageous for all classes of vessels in reducing friction. Eddies are naturally produced in barges having a broad bow and stern ; and the experiments conducted by Mr. de Mas, in 1890–93, show that by making barges spoon-shaped at bow and stern, the resistance due to this cause can be reduced to one-fourth of its ordinary amount, with a diminution in the capacity of the barges of only 4½ to 8½ per cent., according to the type [1]. Fine lines for barges are not needed, and would unduly reduce their capacity for cargo.

## METHODS OF TRACTION.

Several methods of traction have been employed for inland navigation, besides the old system of towage by men and horses. Traction by horses is still largely employed on the canals in England, France, Holland, and Belgium. This method of progression. however, is very slow ; and delays are occasioned when two barges pass one another.

**Steam Traction.** Steamers have long been employed for navigating the larger rivers and canals. Sometimes a tug hauls along a train of barges, as on the Aire and Calder Navigation, where trains of ten to thirty steel coal barges, 20 feet by 16 feet, and 8 feet deep, carrying 35 tons each, are towed by a tug. Frequently steamers carrying cargo are used ; and often these steamers tow one or two barges in addition on large rivers. On very large rivers, where long distances have

---

[1] 'Recherches Expérimentales sur le Matériel de la Batellerie,' F. B. de Mas, Paris, 1891 and 1893 ; and 'Experimental Researches on the Forms of Canal and River Boats,' F. B. de Mas, Water Commerce Congress, Chicago, 1893, p. 19, and plates 3 and 4; and 'Influence de la Forme des Bateaux et l'État de leur Surface sur la Résistance à la Traction,' F. B. de Mas, VIᵐᵉ Congrès International de Navigation intérieure,' The Hague, 1894.

to be traversed, steamers are indispensable for the due development of inland navigation, as on the Amazon, Paranà, Mississippi, Ohio, and St. Lawrence, as well as on the great lakes which are really inland seas. Steam navigation and towage are extensively employed on the Danube, Rhine, Maas, Elbe, Oder, Seine, and other large rivers of Europe, and also on all large ship-canals. On the Rhine, steam-tugs of 1,000 horse-power can tow barges carrying 4,000 tons up-stream ; on the Elbe, between Hamburg and Magdeburg, tugs of 750 horse-power tow a load of 2,750 tons up-stream ; and on the Oder, the maximum load towed by steamers up-stream is 1,500 tons. Paddle-steamers serve best for tugs in rivers of average depth and strong current, and screw-steamers for good depths and a moderate stream. Steamers carrying cargoes of 300 to 500 tons, suited for coasting or short journeys by sea, enable trade to be conveniently conducted between ports along the coast and ports up rivers. Thus steamers of this class ply regularly between French ports on the west coast and Paris, and between London and Paris, and London and Cologne ; and the construction of similar steamers has been proposed for trading between Worcester, on the improved Severn, and the Bristol Channel ports. Several cargo-carrying steamers navigate the French rivers and canals, mainly on the waterways converging to Paris ; and they have an average speed of 4½ to 7½ miles an hour on the Seine, and only 3¾ miles on the canals. Steam-tugs in France are almost wholly confined to the rivers, for they only run regularly on the canals belonging to the municipality of Paris, the Tancarville Canal, and the Marne lateral Canal. In Great Britain, steam is mainly employed on the Aire and Calder and Weaver navigations, the Grand Junction, Bridgewater, Leeds and Liverpool, Warwick and Birmingham, Warwick and Napton, Trent and Mersey, Caledonian, and Grand canals, and on the Thames, Severn, Trent, Lower Avon, and some tidal navigations.

Steam greatly reduces the cost of traction, whilst very materially increasing the speed of transit ; and its advantages are specially experienced on open rivers, over long distances, and with a large traffic. The speed has to be regulated by the size of the waterway, and the condition of its slopes. Steamers undoubtedly provide the most efficient method of traction for general use on waterways ; though other mechanical methods have been resorted to under special conditions, and various systems of traction have been experimented upon.

Haulage on a Submerged Chain. When the current of a river is strong, and the channel is tortuous and shallow in places, there are advantages in laying a chain in sections along the bed of the river, and picking it up by a tug which hauls on the chain by means of a drum on the deck, turned by steam. A chain was laid along forty-five miles of the Lower Seine in 1856 ; and the system was subsequently extended, and is still in use above and below Paris ; and a chain was laid down in the Elbe, in 1869–74, from Hamburg up to Aussig in Bohemia, a distance of 407 miles, by which tugs of light draught were able to haul barges up-stream against a strong current, and over shallows, with much greater facility than the paddle-steamers previously employed, and with only one-third the consumption of coal [1]. When, however, the velocity of the current was reduced, and the depth of water increased, by the canalization of the Seine, and the regulation of the Elbe, improved paddle-steamers were able to compete on more favourable terms with the submerged chain, and reduced their consumption of coal by the introduction of compound engines ; whereas the use of compound engines in the tugs hauling on the submerged chain, produced irregularities in the haulage. Accordingly, ordinary steam-tugs appear to give better results now on the Seine, and

[1] ' La Traction des Bateaux dans les Bassins de l'Elbe et de l'Oder,' E. Bellingrath and Dieckhoff, V^me Congrès International de Navigation intérieure, Paris, 1892, p. 5.

on the Elbe below Torgau. The value of the submerged chain, however, increases with the rate of the current, so that in a river with a fall of 1 in 4,000 to 1 in 3,300, tugs hauling on a chain are as efficient as steam-tugs; and for greater falls than about 1 in 3,300, the chain system becomes the best. Consequently, on the Elbe, the chain and steam-tug work on nearly equal terms between Torgau and the Austrian frontier; but where the fall increases towards the mountains in Austrian territory, the chain possesses so decided an advantage that paddle-steamers have almost ceased to run. Towage by means of a submerged chain is only used on the Rhine between Bonn and Bingen, where the strong current renders it advantageous; it is employed with very satisfactory results on 79 miles of the Neckar; and it has recently been established on the canalized Main, for a distance of 30 miles [1].

Submerged chains have, also, been laid down along the summit-level of three French canals, crossing the water-parting of two river basins in a long tunnel, namely, the Bourgogne, Marne-Rhine, and St. Quentin canals, as well as a reach of the St. Martin Canal, the Ham tunnel of the Canal de l'Est, and a portion of the Deûle Canal, in all of which cases there is a large traffic to be accommodated [2].

The Rhone is a river which, with its strong current and small depth, seems especially suitable for towage up-stream with a submerged chain. The instability, however, of its bed, with large quantities of gravel travelling down in flood-time, has prevented the adoption of a permanently submerged chain, which would be liable to be shifted and buried by the masses of gravel in motion; and, consequently, a steel wire cable has recently been provided, which winding round a drum on the tug, is paid out by the tug when going down-stream,

---

[1] 'Traction des Bateaux sur les Canaux et les Rivières à courant libre du Bassin du Rhin,' Mütze, Vᵐᵉ Congrès International de Navigation intérieure, Paris, 1892, p. 31.

[2] 'Des Meilleurs Modes de Locomotion des Bateaux sur les Canaux,' M. Derôme, Congrès International de l'utilisation des eaux fluviales, Paris, 1889, p. 216.

and is wound up on ascending the river, the cable being laid in sections of about 12½ miles along the river.

Magnetic adhesion has quite recently been introduced, on a new trial tug for towing with a submerged chain on the Seine and Oise, in place of the several turns needed to secure the chain on an ordinary winding drum. These turns strain and wear the chain and render its rapid release from the tug difficult. This new tug, which is 108 feet long and 16½ feet wide, and has a draught of 6¼ feet when towing, has a pulley which can be converted into an electro-magnet at will, and therefore secures the adhesion of the chain in its groove ; and by breaking the electrical circuit, the chain is at once set free, and can be thrown off from the tug, thus dispensing with the drum and the strain on the chain [1]. This tug acts as an ordinary steam-tug in descending the stream, it has been in regular work since 1893, and it has proved so satisfactory that it is proposed to build more tugs of the same type.

**Towing with Endless Chains.** An interesting experiment was made on the Rhone a few years ago, of placing an endless chain on each side of a tug, worked by independent engines, the chains being supported on rollers along the top and resting at the bottom on the bed of the river [2]. The tug was steered by varying the rate of motion of one of the chains ; and the weight of the chain resting on the bottom preventing the slipping back of the tug, the motion was effected against the stream by the rate at which the chains were deposited in front. Arrangements were made for adjusting the length of the submerged chain supported on the rollers, in proportion to the depth, so that the chain might not coil up in shallow water, or have too short a length resting on the bottom in deep water. A portion of the Rhone was

---

[1] 'Traction et Propulsion sur les Canaux, sur les Rivières canalisées, et sur les Rivières à courant libre,' J. Hirsch, VIᵐᵉ Congrès International de Navigation intérieure, The Hague, 1894, p. 12.

[2] Comptes rendus de l'Académie des Sciences, 1883, vol. xcvii. p. 875.

ascended by this vessel at a rate of $2\frac{1}{2}$ miles an hour, where
the velocity of the stream exceeded 7 miles an hour; the fall
was 1 in 1,375; and the depth varied from 5 to 21 feet.
This system, however, does not appear to have received
any practical application.

Haulage by an Endless Travelling Wire Cable. This system
consists of an endless wire cable, supported on grooved wheels,
or pulleys, at a sufficient height above the towing-path, worked
by a fixed motor, and running down-stream along one bank
and up-stream along the other. A barge can attach itself to
the cable at any point by means of a cord with a special clip,
and is hauled along the length of the section, which cannot
be economically made longer than 3 or 4 miles. Changes in
direction are made by guiding the cable at the angle in the
groove of a nearly horizontal wheel; and the speed is limited
to about 2 miles an hour, to enable the tow-rope to be
attached. The essential points to be secured in the system
are, the maintenance of the cable in the grooves of the wheels
in spite of the sideways pull of the tow-rope, the freedom of
the tow-rope from being caught in the grooves, the prevention
of the tow-rope becoming twisted round the cable by the
rotation of the latter, the ready attachment and release of the
tow-rope, and the gradual increase of the pull on the tow-
rope on its attachment to the cable.

The system was first tried by Mr. Riche on the Sambre at
Maubeuge in 1874; and, subsequently, it was experimented on
by Mr. Rigoni, on the St. Martin Canal in 1884, and also on the
Meuse-Scheldt Canal. In 1889, Mr. Oriolle tried his modification of the system on a length of $1\frac{7}{8}$ miles of the St. Quentin
Canal. The method of cable haulage, however, devised by
Mr. Maurice Lévy, and tried in 1888–90 on the St. Maur
and St. Maurice canals, near Paris, is the only one which has
hitherto been deemed capable of a practical application [1].

---

[1] 'Notice Historique et Technique sur le Halage Funiculaire,' Maurice Lévy,
Manchester Inland Navigation Congress, 1890.

The oscillations of the cable, and its tendency to leave the grooves of the wheels when dragged sideways by the tow-rope, have been overcome by making the cable very heavy and giving it a great tension, and also by placing a roller over the summit of the wheel. The tow-rope is prevented from catching in the grooves by side notches on the canal side of the groove; and stop rings are fixed on the cable at definite intervals, to which a stirrup connected with the tow-rope attaches itself when put in a certain position, and from which it can be readily released by a pull on a cord fastened to it. The stirrup also allows of the revolution of the cable without entangling the tow-rope; and the bargeman, by gradually paying out some slack rope at starting, causes a gradual development of the tractive force. The clearing of the tow-rope at concave bends is effected by a slide, which raises the rope so that it clears the groove. The most suitable rate of traction is $1\frac{2}{3}$ to about 2 miles per hour; but this rate, though little more than the rate attained by towing with horses, reduces the time of transit by about one half, owing to the regularity of the traction [1]. The cost of this system of traction varies inversely with the traffic, as the cost of installation and working is little modified by the larger traffic; and the cost has been estimated as ranging between 07 of a penny per ton-mile for a traffic of 1,000,000 tons annually, and ·025 of a penny for a traffic of 3,000,000 tons a year.

This system was established for the first time in a practical form in 1894, for the haulage of barges through the tunnel of Mont-de-Billy, about $1\frac{1}{2}$ miles long, on the summit-level of the Aisne-Marne Canal. This is the sort of case for which the system is specially suitable, as it insures a regular, easy mode of transit through the tunnel, without the smoke of steam-tugs or tugs hauling on a submerged chain, which becomes

---

[1] 'Navigation Works executed in France from 1876 to 1891,' F. Guillain, Transactions of the American Society of Civil Engineers, vol. xxix. p. 24.

very objectionable in a long badly-ventilated tunnel. The system is evidently best adapted for the passage through long tunnels, and along canals where there is a very large traffic requiring to be conducted in a very regular manner.

**Electrical Towage.** Where water-power is readily available, the utilisation of this energy for the generation of electricity to be expended in towage, possesses evident advantages in providing the necessary power in a cheap convenient form, as well as without smoke. This system was established for the first time in 1893, on the summit-level of the Bourgogne Canal, 3¾ miles long, which is in tunnel for rather more than 2 miles of this distance ; and the canal in this reach is only wide enough for a single barge. A discharge of 40,000 cubic yards of water a day, from the watersheds of the Seine and Saône on each side, is available at the two locks at each end of the reach, which turns a Girard turbine at each end, with a combined force of 35 horse-power [1]. These turbines generate the electricity which works the motor on board the tug. The tug is 49 feet long, 10½ feet wide, and 1½ feet draught ; and it hauls on the submerged chain by electro-magnetic adhesion, the chain having been used by the fireless tugs previously employed for the tunnel. The normal rate of progression is about 1½ and 3 miles an hour, according to the gearing used. The total cost of the installation was £5,500 ; and the towage has been carried on very satisfactorily since it was started.

**Towage by aid of Grappling Wheels.** Some steam-tugs navigating the Rhone are provided with a heavy wheel with steel teeth, which grapples the gravel bottom and prevents the tug slipping back when ascending the river. Where, however, the river-bed is soft, or very hard, or the river is deep, the grappling wheel does not work satisfactorily.

---

[1] 'Traction et Propulsion sur les Canaux, sur les Rivieres canalisées, et sur les Rivières à courant libre, J. Hirsch, VIᵐᵉ Congrès International de Navigation intérieure, The Hague, 1894, p. 5.

Nevertheless, the ordinary speed attained by these tugs with this arrangement is about 3 miles against the current.

**Towing by Locomotives on the Bank.** Towage was carried on for many years by locomotives on the Neuffossé, Aire, and Deûle canals in France, along a length of 48 miles, with only one intervening lock[1]. The locomotives ran on a metre gauge laid with rails weighing 30 lbs. per lineal yard, and they weighed 14 tons when loaded. They towed two or three fully loaded barges up-stream, at a speed of about a mile an hour; but the charge for towage was fixed too low, so eventually this towage was discontinued.

Locomotives were subsequently tried for towing barges in Germany; but, though the system worked satisfactorily, it was not economically successful. Moreover, this method of towage, though it could be worked at a speed of 4 to 8 miles an hour, would not be applicable on a canal with frequent locks or sharp bends. The system was also experimented upon for towing barges on the Shropshire Canal; but, though the towing was satisfactorily accomplished, difficulties were encountered of a practical nature, which led to its abandonment.

**Remarks on Methods of Traction.** Besides the long-established methods of traction on waterways, of steam-tugs and haulage on a submerged chain, only three systems have been hitherto adopted for practical purposes. Grappling wheels have been used for a long time on the Rhone; but the system is somewhat defective, and will probably be superseded by the new movable cable haulage, if it affords good practical results. Endless travelling cables may prove advantageous under special conditions for a large traffic; but they are unlikely to receive a very general extension. Electrical towage has proved very successful with an abundant supply of water available, and where the absence of smoke is important; and it is probable that under similar special conditions, the system

---

[1] 'Des Meilleurs Modes de Locomotion des Bateaux sur les Canaux, M. Derôme, Congrès International de l'utilisation des eaux fluviales, Paris, 1889, p. 218.

may be more generally adopted. Much attention is being devoted to various systems of traction ; and trials of the systems should result in useful information as to their relative economy and practicability.

## CLASSIFICATION OF SHIP-CANALS.

Ship-canals may be divided, for convenience of description, into three categories, namely, (1) Ship-canals providing a navigable channel for vessels, in place of a portion of a river ; (2) Ship-canals affording ports with a shallow approach, or inland towns direct access to deep water in the sea ; and (3) Ship-canals connecting two seas, by cutting across the intervening land which generally constitutes an isthmus. The first of these classes of ship-canals will be considered in this chapter, reserving the two remaining classes for the two concluding chapters of this book.

Sometimes a large tidal river is so inconvenient for navigation along a portion of its course that a lateral ship-canal is advantageously provided in its place, as for instance in the case of the tidal Severn between Gloucester and Sharpness, and the tidal Loire between La Martinière and Carnet Island. Occasionally, a tidal river, in flat country, follows such a circuitous course that a ship-canal furnishes a much shorter route, as well as an opportunity of affording a better depth, as in the case of Ghent's access to the sea, which is very long by the Scheldt. The shallow outlets of tideless rivers have in some cases been avoided by the construction of a ship-canal, diverging from the river near its mouth, and entering the sea at a point where the depth has not been reduced by the advance of the delta. Thus, besides the instances and proposals referred to on p. 199, the St. Louis Ship-Canal has been formed to avoid the Rhone bar, and the St. Petersburg and Cronstadt Ship-Canal has been dredged to provide a deeper outlet into the Baltic than the channel of the Neva affords across the shallow foreshore, as well as the

Fiumicino Canal (p. 499), which gives small vessels access to the Tiber away from the obstructions presented by its delta.

## LATERAL SHIP-CANALS.

Generally a large tidal river is capable of adequate improvement for navigation by training walls, regulation works, or dredging; but in one or two instances a lateral ship-canal has been resorted to in preference, serving the same purpose for sea-going vessels as the Welland, St. Lawrence, and other large lateral canals do for inland navigation.

**Gloucester and Berkeley Ship-Canal.** The Severn for several miles below Gloucester is so tortuous, and the tidal currents are so rapid, owing to the great range of tide, that the navigation is difficult and dangerous. Accordingly, as early as 1793, an Act was obtained for connecting the Severn at Gloucester with its estuary at Sharpness, in the parish of Berkeley, by a fairly direct ship-canal, to enable vessels to avoid that section of the river, and shorten their route. The canal, however, was only completed by Telford in 1827, at a cost of £500,000 ; its length is 16½ miles, in one level reach, as compared with 26 miles by the river. A lock at Gloucester gives the canal access to the river with its varying water-level ; and a tidal lock at the lower end of the canal, opening into the estuary under the shelter of Sharpness Point, connects the canal with the tidal estuary.

The canal was made 80 to 100 feet wide at the water-level, and 13 to 20 feet at the bottom, and wider at passing places ; and its depth of water was 18 feet, so that at the time of its completion it was rightly regarded as a ship-canal (Plate 11, Fig. 18). The size, however, and draught of ocean-going vessels have greatly increased since that period ; whilst the depth of water in the canal has shoaled to 15 feet. Access, however, to the canal from the estuary has been improved by the construction of a tidal basin, with an entrance into the estuary a little lower down, having a width of 60 feet and a depth over the sills of 16 feet at neaps and 29 feet at springs, and with a lock opening into a dock leading to the canal, 320 feet long and 60 feet wide. The lock at Gloucester, which formerly was only available for vessels of a maximum draught of 7¾ feet, has recently been deepened so as to admit vessels drawing 9½ feet ; and whilst vessels of about 600 tons can go up the canal to Gloucester, vessels of 300 tons can now pass up the Severn to Worcester ; and the traffic on the canal has increased considerably in recent years.

Part of the scheme for improving the water communication between Birmingham and the Bristol Channel, consists in straightening the bends of the canal, and dredging it to its original depth, so as to enable steamers 230 feet long and 16 feet draught to get up to Gloucester. Moreover, as sea-going vessels of large draught cannot reach the dock at Sharpness at neap tides, the construction of a new entrance at Sheperdine, 5½ miles lower down the estuary, has been proposed, providing 8 feet greater depth at neap tides, and connected with Sharpness by a ship-canal 135 feet wide and 25 feet deep[1]. The cost of these proposed works, between Sheperdine and Gloucester, has been estimated at £330,000.

This ship-canal has not only enabled a small class of sea-going vessels to navigate safely up to Gloucester, but it has also led to the recent improvement of the Severn up to Worcester ; and it, moreover, renders the scheme for connecting Birmingham with the sea by an adequate waterway far more practicable, by having enabled a waterway accessible to vessels of 300 tons to be brought within thirty miles of Birmingham[2].

**Tidal Loire Ship-Canal.** The training works in the Loire estuary, between Nantes and La Martinière, produced such serious accretions in the estuary below (pp. 306–308), that instead of prolonging the training works, it was decided, in

---

[1] 'Birmingham and Bristol Channel improved Navigation,' Gloucester, 1890.

[2] 'Report of the Proceedings of the Conference on Inland Navigation at Birmingham, 1895,' p. 106, and plate 26, fig. 3.

1879, to construct a lateral ship-canal along the left bank of
the estuary, between La Martinière and Carnet Island (Plate 9,
Figs. 4 and 5). This ship-canal, 9⅓ miles long, with a bottom
width of 78¾ feet and a normal depth of water of 19⅔ feet,
has access to the river at each extremity through a lock, 558
feet long and 59 feet wide, worked by hydraulic machinery, to
provide for the varying heights of the river ; and it was completed in 1892. It enables large vessels navigating between
Nantes and the sea, to pass, in a deep sheltered channel of
uniform depth, between the outlet of the trained channel at
La Martinière and the deep channel situated between Carnet
Island and the left bank of the estuary, and thus to avoid the
shallow circuitous shifting channel of that section of the
estuary.

**Tancarville Canal.** The canal between Tancarville and
Havre, formed through the alluvial deposits along the right
bank of the Seine estuary (p. 317), at present only serves as
a sheltered channel, 15½ miles long, by which the river craft
can pass from the trained channel of the Seine to Havre,
without entering the open estuary (Plate 9, Fig. 1). The
canal, however, has been so constructed that it will be only
necessary to dredge the bottom between Tancarville and the
Harfleur branch canal, so as to increase the depth from 11½ to
19⅔ feet, in order to convert it into a lateral ship-canal for sea-
going vessels, in the event of the navigable channel of the
estuary, between Berville and the open sea, becoming so deteriorated by gradual accretions as to be incapable of being used
by large vessels. The locks at both ends have, accordingly, been
made 52½ feet wide at their entrances, leading into a chamber
590 feet long and 98½ feet wide ; and the depth of water over
the inner sills is 19⅔ feet at the normal water-level of the
canal, and 23 feet over the outer sills at high water of a low
neap tide[1]. The swing-bridges also for roads across the

---

[1] 'Canal de Havre à Tancarville,' Maurice Widmer, Havre, 1887 ; and Annales
des Ponts et Chaussées, 1892, (1) p. 633, and plates 7 to 20.

canal have been given the full span necessary for a ship-canal; and the canal has been carried to its full width throughout of 136 feet at the water-level, exclusive of the berms, between Havre and the Harfleur branch (Plate 11, Fig. 22), and 143 feet between the Harfleur branch and Tancarville, with a present bottom width of 82 feet and side slopes of 3 to 1. Whilst, however, the depth of the former portion has been carried to the full extent of 19¾ feet, the latter and much longer portion has been only made 11½ feet deep, which is sufficient for the present purpose of the canal.

The excavation of the canal was commenced by excavators, which were served by endless india-rubber travelling bands, 3¼ feet wide, carried by lattice girders, 177 feet long, for conveying the material to the banks. The cutting for the canal was completed, on the admission of water into the trench, by bucket-ladder dredgers provided with long shoots for depositing the dredged material at the sides.

**Ghent-Terneuzen Ship Canal.** The important position held by Ghent in the Middle Ages, led its inhabitants to endeavour to form a better outlet for its trade than the very circuitous channel of the Scheldt. As early as the middle of the thirteenth century, a canal was excavated from Ghent in the silted-up bed of the Lième, with an outlet at Damme, the port of Bruges; and in the sixteenth century a direct route was obtained to the estuary of the Scheldt, by the construction of a canal from Ghent to Sas-de-Gand, where it entered a branch of the Brakman, a tributary of the Scheldt. This route, which was opened in 1561, was closed in 1648 by the treaty of Munster, and remained closed till 1815; and another route was only provided in 1758, by the opening of the Ghent-Bruges Canal, which, however, placed Ghent at a great disadvantage in respect to its ancient rival Bruges. The Brakman was in the meantime silting up; but at last a new outlet was provided by the extension, in 1825–27, of the old canal from

Sas-de-Gand to Terneuzen, where the deep navigable channel of the Scheldt estuary runs close alongside the shore[1]. This canal served at first both for navigation and drainage. The discharge, however, of the water in flood-time at Terneuzen caused abrupt variations in the water-level and rapid currents; and, accordingly, drainage canals were constructed along both sides of the lower reach of the canal in 1844–45, which diverted the flood water below Sas-de-Gand; and further drainage improvements have rendered the discharge of flood waters at Terneuzen very exceptional since 1886.

The canal was given a minimum depth of 13 feet from Ghent to Terneuzen; and the reconstructed lock at Sas-de-Gand divided the canal into two reaches of $12\frac{2}{5}$ miles and $8\frac{3}{10}$ miles respectively, with a difference in level of $1\frac{1}{2}$ feet. The lock at Sas-de-Gand was made $39\frac{1}{4}$ feet wide, with a depth of $14\frac{1}{4}$ feet over the sills; and the tidal lock at Terneuzen was built $26\frac{1}{4}$ feet wide, with an available depth of $19\frac{1}{4}$ feet over the sills at mean high water. Screw steamers were first allowed to navigate the canal in 1851.

In 1879, Holland and Belgium entered into an agreement for the enlargement of the canal, which has been given a depth of $21\frac{1}{3}$ feet in Belgium and about 20 feet in Holland, a bottom width of $55\frac{3}{4}$ feet, and side slopes of 3 to 1 (Plate 11, Fig. 23); and the bends of the canal have been made easier. New locks, also, have been constructed in side cuts at Sas-de-Gand and Terneuzen, 295 feet long and $39\frac{1}{3}$ feet wide, with depths of $20\frac{2}{3}$ and 19 feet respectively over their sills. Ghent has therefore been provided with a ship-canal shortening the route to the Scheldt estuary at Terneuzen, from 105 miles by the Scheldt to $20\frac{7}{10}$ miles by the ship-canal, and providing a constant minimum depth of 19 feet, in place of a variable depth by the tidal Scheldt, which at low tide falls to $6\frac{1}{4}$ feet, and does not exceed $17\frac{3}{4}$ feet at high tide.

[1] 'Les Voies de Navigation dans le Royaume des Pays-Bas,' The Hague, 1890, p. 241.

The traffic by this route passing Sas-de-Gand has increased enormously in the last forty years, the increase being about fivefold for the inland traffic, and about fourteenfold for the maritime traffic; and whilst the growth of inland traffic has fluctuated since 1882, the maritime traffic has more than doubled since 1882. The recent commercial development of Ghent is in great measure due to the enlargement of this canal, which has enabled it to obtain a traffic by water about equal to, and likely to exceed that of Brussels, so that Ghent, amongst Belgian ports, is only overshadowed by Antwerp, whose rise as a port during recent years has been exceptionally rapid.

St. Louis Ship-Canal. A canal was constructed in 1802, leaving the Rhone at Arles and entering the Mediterranean at Bouc on the north-eastern side of the Gulf of Foz, in order to provide an outlet for the traffic of the Rhone in deep water beyond the zone of the deposits of the Rhone delta. As, however, this canal was given a depth of water of only 6½ feet, and a width of 26¼ feet at the locks, it was too small to accommodate the traffic directly steamers were introduced on the Rhone.

The St. Louis Ship-Canal was constructed in 1863–73, in order to provide a deep-water outlet from the Rhone, which the jetty works, carried out in 1852–57, and concentrating the whole discharge of the Great Rhone into the eastern mouth, had failed to form across the bar outside (p. 188). This canal starts from the left bank of the Rhone at the town of St. Louis, only about 4 miles above the ends of the jetties, and runs nearly due east into the Gulf of Foz[1]. It is slightly over 2 miles long, and has a depth of 19⅔ feet at the lowest sea-level; its width at the water-level is 207 feet; and its bottom is 98½ feet wide across the land, 197 feet wide from the shore out to the 13-foot line of soundings, and 656 feet

---

[1] Minutes of Proceedings Institution C. E., vol. lxxxii, p. 316, and plate 6, figs. 5 and 6.

wide from this depth out to the 19⅝-foot line. The turbid waters of the Rhone are shut off from the canal by a lock at the entrance, pointing down-stream so as to be protected from the current of the river. The lock has an available length of 525 feet, a width of 72 feet, and a depth of 24½ feet of water over the sills. The lower end of the canal is quite open to the sea, but well sheltered by the bay; and though it is within the zone of the deposits from the river, the accretion in ten years was only about 4 inches. The advance, however, of the delta eastwards, across the entrance to the gulf, is rendering the approach to the canal from the west more circuitous, and already impedes access to it during northerly winds, which may necessitate the reopening of the southern outlets of the Rhone.

The ship-canal provides a much better depth than the improved Lower Rhone generally affords ; whilst a basin of 30 acres between the lock and the canal, provides a sheltered place for the transhipment of cargoes. The prospects of the canal have been increased by the improvement of the Rhone below Lyons, and the gradual growth of the traffic on the river ; for the traffic on the canal, which was quite insignificant till 1881, when its tonnage was only 29,350 tons, rose regularly each year up to 271,000 tons in 1891, since which year the canal has shared in the general depression of trade. The canal depends wholly on the traffic on the Lower Rhone, which remains small in proportion to the length and size of the river ; a portion of this traffic goes by the Cette and Bouc canals ; and the traffic is undoubtedly hindered by the necessity of transhipment between the sea and river craft at St. Louis, and the absence of any local trade. The St. Louis Ship-Canal, accordingly, though it has fully achieved the object for which it was constructed, of providing a deep-water outlet for the Rhone, has not hitherto become an important waterway owing to unfavourable conditions.

St. Petersburg and Cronstadt Ship-Canal. Though the waters discharged by the Neva into the Gulf of Finland at

St. Petersburg are deprived of their alluvium in flowing through Lake Ladoga, the sea in front of the mouth of the Neva is very shallow and obstructed by sandbanks, so that the navigable channel as far as Cronstadt was only between 7 and 20 feet deep ; and the cargoes of vessels trading with St. Petersburg had to be transhipped at Cronstadt. There appears also to be a gradual rising of the sea-coast at St. Petersburg, increasing the area of dry land round the city, which may account for the shallowness of the foreshore through which the outlet channel of the Neva flows to deep water at Cronstadt. Peter the Great commenced a canal for providing better access to St. Petersburg, but the works were abandoned at his death.

The ship-canal, commenced in 1877 and completed in 1885, starts from the Neva below St. Petersburg, and diverging from the outlet channel in the estuary, it goes to the south-west for 2 miles, and then curving round to north-west, proceeds in a straight line to Cronstadt [1]. The canal has been dredged to a depth of 22 feet throughout its whole length of 17½ miles ; for the first 7½ miles it is protected by embankments on both sides ; and it has been given a bottom width of 207 feet to the end of the curve, and 275 feet beyond. The embankments were formed with some of the excavated material, deposited between two rows of fascines and protected by sheet piling and planking up to low-water level in the shallower parts, and pitched with shingle on the outer slope. In the deeper portions, the embankment was formed between two rows of timber caissons, the lower portion being deposited from hopper barges ; and the embankment was then raised to the full height by the silt discharged from sand-pump dredgers, and the slopes pitched. The last ten miles of the canal were merely dredged through the shallow gulf, and no embankments were formed.

[1] 'Atlas des Ports de Commerce de la Russie,' part 1, 1892, plates 1 and 2 ; and Mémoires de la Société des Ingénieurs Civils, Paris, 1883, (1) p. 512, and plates 50 and 51.

The canal has provided an excellent navigable channel between deep water in the Baltic and St. Petersburg, 22 feet deep, at a cost of about £1,240,000, so that sea-going vessels can get up to St. Petersburg, where three basins, with an area of 430 acres, have been formed for their accommodation by widening the canal at places; and the canal, which was constructed by the government, is free of toll.

# CHAPTER XXIII.

## SHIP-CANALS FOR PORTS.

SHIP-CANALS constructed to important towns which are either inland, or possess only a shallow, or very circuitous access to the sea, will be considered in this chapter. Such canals serve to convert an inland town into a seaport, as in the case of the Manchester Ship-Canal; or they provide a deep-water access to the open sea for old ports which, owing to the increased draught of vessels, have too shallow an approach to the sea, of which the North Holland and Amsterdam ship-canals are examples; or they secure a direct and safe channel to the open sea for ports on an inland sea, which may be of great importance for strategical objects, as illustrated by the recently opened Baltic Canal. These four ship-canals, three of which are illustrated on Plate 12, exhibit considerable differences in the conditions under which they had to be designed; and whilst three of them have been formed to provide direct deep-water access to the open sea for ports on inland seas, in one level reach, the fourth, besides having a long tidal reach, rises by locks to an inland city, and has converted it into a seaport.

## NORTH HOLLAND SHIP-CANAL.

**Object of the Canal.** The Zuider Zee formerly provided the only means of access to the port of Amsterdam; and owing to the continual shoaling of the shallow approach channel, which had an available depth of only 11½ feet about the year 1825, Amsterdam was in danger of losing her commerce, when the cessation of the continental wars in 1815 created a revival in the ocean-going trade, in view of her unfavourable situation in comparison to other seaports. Accordingly, a deeper approach to Amsterdam was urgently needed to maintain her trade; and as the direct western route to the North Sea was considered impracticable, on account of the necessity of maintaining the drainage of the canals flowing into Lake Y, and the difficulties anticipated in forming and maintaining a deep entrance on the flat sandy shore of the North Sea, a northerly course was selected for a canal to connect Amsterdam with the existing deep sheltered harbour of Niewediep, adjoining the Texel Roads. This canal, traversing North Holland, was constructed in 1817–24, at a cost of over £916,000; and it was a remarkable work considering the period of its execution, and the alluvial foundations on which the locks at each end had to be built.

**Description of the North Holland Canal.** This canal, starting from Lake Y, opposite Amsterdam, and proceeding northwards to Niewediep by a somewhat tortuous route, necessitated probably by local conditions, is 50 miles long;

and it has a width of about 124½ feet at the water-level and 33 feet at the bottom, and a minimum depth of water of 18 feet (Plate 11, Fig. 24). Besides a tidal lock at each end of the canal, with gates pointing seawards to keep out the tide, three intermediate locks were built for regulating the water-level at different parts of the canal, and to assist the tidal locks in preventing the influx of high tides into the canal [1]. The locks were made double, with a large and small lock for the accommodation of vessels of different sizes, the larger lock having been given an entrance width of 51½ feet, a depth over the sills of 21½ feet, and a length of chamber of 213¼ feet. Seventeen opening and floating bridges were provided for roads across the canal, leaving a clear width of 51½ feet when open, which has been increased on reconstructing the bridges to a minimum of 58 feet. The locks also at Niewediep, and opposite Amsterdam, were rebuilt in 1851–55 and 1861–65, with lengths of 228 and 358 feet, widths of 55¾ and 59¾ feet, and depths over the sills of 21⅓ and 24½ feet respectively. The canal also has been straightened at some of the worst curves, and its width increased.

**Condition of the North Holland Canal.** Though the canal rendered great services to the trade of Amsterdam for many years, its moderate section and small locks proved inadequate, and its curves too inconvenient, for the increasing size and draught of sea-going vessels. A proposed enlargement of the canal, to suit modern requirements, was abandoned on account of the length and circuitous course of the canal to the sea, involving serious loss of time and additional cost in transit, important considerations in the face of the keen competition of other ports more favourably situated by nature, which could only be remedied by adopting a direct route to the North Sea.

Since the opening of the Amsterdam Ship-Canal, sea-going

---

[1] 'Les Voies de Navigation dans la Royaume des Pays-Bas,' The Hague, 1890, p. 26.

merchant vessels rarely use the North Holland Canal, which, however, still serves for inland navigation, and is valuable for vessels of war. The chief traffic takes place at the Amsterdam end of the canal.

## AMSTERDAM SHIP-CANAL.

The cutting of a canal across the narrow strip of land separating the North Sea from Lake Y and Wyker Meer, was first proposed in 1629, mainly for the better drainage of the low-lying polders, the accommodation of shipping forming quite a secondary consideration.

**Purpose of the Canal.** Though a canal providing direct access from Amsterdam to the North Sea was rejected as impracticable in 1817, the increasing necessity of obtaining a short deep-water access for Amsterdam, led to a number of projects being proposed between 1852 and 1863, for forming a direct canal to the North Sea. At last, in 1863, a concession was granted for the construction of the canal, which was transferred in 1865 to the Amsterdam Ship-Canal Company, who proceeded to execute the work [1].

In place of the circuitous, inadequate channel of the North Holland Canal, 50 miles in length, opening only into the Texel Roads, the Amsterdam Ship-Canal affords a direct deep channel from Amsterdam to the North Sea, only about 16 miles long, entering the sea at a convenient point for trading with England and the various continental ports (Plate 12, Fig. 1).

**Difficulties of the Scheme, and Methods of surmounting them.** Some difficulties at once presented themselves in the apparently simple scheme of connecting Wyker Meer with the North Sea, by a cut across the intervening strip of land, only about 3¾ miles in width, the two principal of which had led to the northern route being preferred in 1817, as mentioned

, [1] 'Les Voies de Navigation dans le Royaume des Pays-Bas,' The Hague, 1890, p. 22.

above. Besides the necessity of providing a permanent, deep outlet across the exposed, shallow, sandy foreshore of the North Sea, and of maintaining the drainage of the adjacent low-lying lands, the water communication of the villages bordering the lakes, with each other and Amsterdam, had to be preserved, when the proposed reclamation of the lakes was carried out. The varying levels, moreover, of the Zuider Zee which joined Lake Y, and had to be placed in communication with the canal, would have proved very inconvenient in the canal.

The formation and maintenance of a deep, sheltered outlet for the canal on the North Sea coast did not present the same difficulties in 1865 as in 1817, owing to the experience gained in harbour works since that period, and the improvements effected in dredging machinery. In reclaiming the lakes on each side of the canal, the communications of the villages were preserved by forming branch canals from them leading direct to the ship-canal, which also provided channels for the passage of the drainage waters from the low-lying lands into the ship-canal, which formerly drained into the lakes. The oscillations in level of the Zuider Zee were excluded from the canal, and a sufficient fall for the drainage of the lands secured, by shutting off the Zuider Zee from the canal by an embankment across Lake Y, 2 miles to the east of Amsterdam, and by keeping the water in the canal permanently down to only 14 inches above ordinary low-water level in the North Sea by letting out as much of the surplus drainage water as practicable at low tide through sluices into the sea, and by discharging the greater portion into the Zuider Zee by pumps erected at the northern end of the embankment[1]. Locks at each end provide communication with the North Sea, and across the bank at the Zuider Zee end ; and reverse gates pointing seawards prevent the influx

[1] 'The Amsterdam Ship-Canal,' Harrison Hayter, Minutes of Proceedings Institution C.E., 1880, vol. lxii, p. 3, and plates 1 to 3.

of water from the sea at either end into the canal, and secure the low-lying lands from inundation (Plate 12, Figs. 1 to 5). The rise of ordinary tides is about 5½ feet in the North Sea at Ymuiden, and only 1¼ feet in the Zuider Zee near Amsterdam, whose level is more affected by wind.

**Description of the Amsterdam Ship-Canal.** The canal between Amsterdam and the North Sea has been made as straight as practicable, consistent with passing through Wyker Meer, and giving a suitable course for the reclamation of the lakes, and for avoiding unnecessary excavation (Plate 12, Fig. 1). The canal was originally formed with a bottom width of 88½ feet, a depth of water of 23 feet, side slopes of 2 to 1, a small berm at the water-level through the sand hills, and a much wider berm a little below the water-level through the lakes, giving a width of 176 feet at the water-level through the sand hills, and 360 feet through the lakes including the berm (Plate 11, Fig. 25). A berm at, or above the water-level in a deep cutting, preserves the canal from the falling in of material on the occurrence of slips; and a berm below the water level protects the soft side banks, formed of dredged material, from the waves raised by passing steamers, which are expended in spreading over the wide berm.

The canal was first excavated through the sand hills at the North Sea end; and the material was used to form the core of the banks, on each side of the canal, through the lakes. The maximum depth of cutting was 95 feet; and the sides were protected near the water-level by piles with planks and sheeting, and broken bricks at the back. As the banks progressed. the canal was dredged between them through the shallow lakes, the dredged material being deposited on the banks through floating tubes, as previously described (p. 94). Dredging was employed for forming the canal for about 10½ miles through the lakes, and extending it into deep water across the foreshore of the North Sea; but on approaching Amsterdam, Lake Y is sufficiently deep (Plate 12, Fig. 2).

The banks were protected by fascines, and the land reclaimed behind them; and the drainage of the 14,330 acres of reclaimed land is effected by pumping the water into the canal, or into the branch canals, which were formed at the same time, of shallower depth, to connect the villages with the canal. The total quantity of material excavated and dredged for forming the canal amounted to 21,000,000 cubic yards.

The canal is in communication with the North Holland Canal, the Merwede Canal, and the other inland canals converging to Amsterdam; and it provides access to the ports and canals opening into the Zuider Zee through the Orange, or Zuider Zee locks. The portion of the canal between the Ymuiden, or North Sea locks and Ymuiden Harbour, about 7 furlongs long, forms a sort of sheltered, inner tidal harbour for vessels waiting to pass through the locks or to go out to sea; and it has, consequently, been widened to 115 feet at the bottom, and deepened to 28⅔ feet below ordinary low water. The canal is crossed by one roadway swing-bridge near Velsen, and two railway swing-bridges, leaving a clear waterway when open of 62 to 63 feet. The length of the reach of the canal between the locks is 17½ miles; and seven branch canals, of different lengths and sizes, lead straight out of the ship-canal to the villages on the old shore of Lake Y (Plate 12, Fig. 1). The works were commenced in 1865, and the canal was opened in 1876.

**Zuider Zee Embankment and Locks.** The embankment shutting off the Zuider Zee from the canal, 4,462 feet long, was constructed in 1866–72; it was formed of fascine mattresses at the bottom and sides, filled in with clay and sand in the middle; and it was protected on the top and slopes by a layer of 3¼ feet of puddled clay covered by stone pitching, to secure it against waves (Plate 12, Fig. 5).

The Zuider Zee locks were constructed at the same time within a circular cofferdam, 525 feet in diameter, in Lake Y, in the line of the embankment, as the locks had to be completed before the embankment could be closed, so as not to stop the traffic between the Zuider Zee and Amsterdam. There is one large lock in the middle, and a smaller lock on each side; beyond which there are three passages to the north for discharging the water raised by the pumps from the canal, and a sluiceway to the south for the discharge of the drainage water from the canal, on the rare occasions when the level of the Zuider Zee is low enough (Plate 12, Fig. 4). The large lock has a chamber 315 feet long and 59 feet available width ; and the sills of all three locks are 14½ feet below low water. These locks have to provide for the passage of the previously existing traffic between Amsterdam and the ports on the Zuider Zee, as well as the traffic between the North Sea and the Zuider Zee; but only vessels of moderate draught have to be accommodated at these locks, owing to the shallowness of the channels of the Zuider Zee.

Every lock, at both ends of the canal, is furnished with an intermediate pair of gates pointing seawards to expedite the lockage of small vessels down into the canal ; and all the locks and sluiceways have a pair of sea-gates at each end, which, together with the intermediate pair, effectually secure the canal, and the adjacent low-lying lands, from any chance of the influx of the sea at either end (Plate 12, Figs. 3 and 4). The gates pointing towards the canal at each end of the locks are made of wood, and those pointing seawards of iron to ensure greater security from the sea.

The pumping machinery at the northern end of the Zuider Zee embankment consists of three centrifugal pumps, with fans 8 feet in diameter, worked by a steam-engine of 200 horse-power. Another steam-engine, of 300 horse-power, has recently been erected at the southern side of the locks, to turn six scoop wheels, 28 feet in diameter and 10 feet wide, for removing more promptly the drainage water flowing into the canal from an extensive area, including most of the land reclaimed from Haarlem Meer, amounting sometimes to between 5,000,000 and

9,000,000 cubic yards in 24 hours, and thus keep down the water-level of the canal more uniformly at the intended level.

**Ymuiden Locks.** As the vessels passing between the North Sea and Amsterdam are less numerous, but for the most part considerably larger than those passing through the Zuider Zee locks, only two locks were constructed at Ymuiden, together with a sluiceway for discharging the drainage water into the North Sea by gravitation when possible (Plate 12, Fig. 3). The large lock has an available length of 393 feet, and a width of 59 feet; and its sills are 23⅓ feet below low water, to admit the vessels of large draught trading with Amsterdam. The smaller lock, with a length of 229 feet, a width of 39⅔ feet, and a depth over the sills at low tide of 14 feet, is used for the discharge of water from the canal in flood-time, as well as for navigation.

As the large lock has already become inadequate to admit the large ocean steamers of the present day, a new single lock has been constructed in a wide and deep side-cut parallel to the canal on the north side, which was commenced in 1888 (Plate 12, Figs. 1 and 3)[1]. The lock has an available length of 740 feet, a width of 82 feet, and a depth over the sills at low tide of 30⅓ feet; and it has three pairs of sea gates adjoining the three pairs of gates pointing towards the canal. The cost of this lock has been estimated at £458,000.

**Ymuiden Harbour.** The canal is protected at its outlet into the North Sea by two moles projecting, on each side, into Ymuiden Harbour, and formed of fascines covered by rubble stone. The harbour in the North Sea is formed by two converging breakwaters, each 5,013 feet long, built of concrete blocks placed upon a foundation layer of basalt, and capped with concrete in mass (Plate 12, Figs. 1 and 6). The layer of basalt, 3¼ feet thick, which extends 33 feet beyond the base of the breakwaters on the sea side, was introduced because

---

[1] 'Guide du VIᵐᵉ Congrès International de Navigation intérieure,' The Hague, 1894, p. 17.

it was found that the scour of the sea along the face of the break-water washed the sand from under the blocks, which were at first laid on the sandy bottom. The damage, also, sustained by the breakwaters, rendered it necessary to protect the sea face of the outer half of each breakwater, and the pierheads, with a wave-breaker of random concrete blocks. As the blocks forming the exposed surface of the wave-breaker were subject to disturbance in storms, their minimum weight was subsequently fixed at 20 tons, and the proportion of Portland cement was doubled ; and as they still were liable to be shifted and fractured, the additions to the mound are being laid in regular rows like steps, and the blocks at the outer ends have in some cases been chained together. The concrete-in-mass capping of the breakwater has also been raised, and the top paved with hard bricks laid in cement; but the repairs of the breakwaters and wave-breakers still necessitate a con-siderable annual expenditure.

The breakwaters, which are about 3,940 feet apart at the shore, converge seawards so as to leave an opening of only 853 feet between the pierheads ; and the available navigable entrance is reduced by the wave-breakers to 656 feet. The sheltered area enclosed by the breakwater is 247 acres, through which a channel has been dredged out to deep water, with a bottom width of 490 feet where the depth at low water reaches $28\frac{1}{2}$ feet, and 820 feet at the minimum depth of 25 feet. The maintenance of this channel, both inside and outside the harbour, necessitates the removal, by sand-pump dredgers, of 650,000 cubic yards, on the average, in a year.

**Enlargement of the Amsterdam Ship-Canal** Soon after the opening of the canal, it was recognised that the depth determined on in 1865 was insufficient for the growing re-quirements of vessels; and the depth was increased, in 1877–1883, from 23 feet to $25\frac{1}{4}$ feet, with a bottom width of 65 feet. In 1889, a further deepening of the canal was commenced at the North Sea end, which is being gradually extended towards

Amsterdam, and is to be completed in 1897. By these works, the canal is being made 28 feet deep up to Amsterdam; and the bottom width is being increased to 82 feet along the ordinary section, and 105 feet at the sidings, which will altogether amount to one-fourth of the length of the canal; and the width at the water-level will be generally from 328 to 426 feet[1] (Plate 11, Fig. 25).

The largest vessels that have hitherto been admitted into the canal, through the Ymuiden locks, are 377 feet long, 58¼ feet beam, and 23⅗ feet draught. On the completion of the new lock and the enlargement works, vessels 567 feet long, 51 feet beam, and 26¼ feet draught, will be able to navigate the canal up to Amsterdam; whilst the new lock at Ymuiden could give passage to vessels of the future, 722 feet long, 65⅗ feet wide, and 30½ feet draught; but a very large expenditure would be necessary to enlarge the canal sufficiently to accommodate vessels of these dimensions[2].

**Cost of the Amsterdam Ship-Canal.** The canal was ceded by the Company to the State in 1882; and the expenses up to that period have been reckoned at £4,853,000. The cost, however, of the original works has been estimated at only £3,000,000; and the reclamation of the 14,330 acres from Lake Y and Wyker,Meer, realised £1,166,000 by the sale of the reclaimed lands.

**Traffic on the Amsterdam Ship-Canal.** The traffic passing through the locks at Ymuiden increased steadily from 1877 to 1892, being fivefold in the latter year what it was in the former; and though, in common with most of the Dutch waterways, there has been a slight diminution since 1892, it is probable that the completion of the enlargement of the canal will promote the traffic, by affording access for larger

---

[1] 'Guide du VIᵐᵉ Congrès International de Navigation intérieure,' The Hague, 1894, p. 18.

[2] 'The Enlargement and Improvement of the North Sea Canal of Holland,' A. E. Kempers, Transactions of the American Society of Civil Engineers, vol. xxx, p. 405, and plates 5 and 6.

vessels. The growth of traffic along the canal is, indeed, considerably less than along the Maas up to Rotterdam ; but the canal has undoubtedly enabled Amsterdam to maintain and improve her position amongst commercial ports, and the improvements in progress will tend to increase the trade of Amsterdam.

## BALTIC SHIP-CANAL.

The idea of forming a waterway across a narrow part of Holstein, uniting the Baltic by a direct route to the North Sea, and thereby avoiding the circuituous, difficult vogage round Jutland, appears to have originated some centuries ago. This idea was partially realised by the construction, in 1777–1784, of the Eider Canal, 26 miles long, which connected the Baltic at Holtenau, near Kiel, with the river Eider flowing into the North Sea, and enabled vessels of 120 tons, by a transit of about 100 miles along the canal and the Eider, to pass between the Baltic and the North Sea. As this canal rose 23 feet by three locks to a summit-level, and descended by three locks to the Eider which is shallow and winding, the enlargement and rectification of this canal throughout, to provide a route for sea-going vessels of the present day, would not have satisfied the requirements of Germany. The Baltic Canal, accordingly, whilst following approximately the line of the Eider Canal between Kiel Harbour and the Eider Lakes, but with a much straighter course, does not utilise the river Eider beyond, though following its valley ; it diverges entirely from the course of the river at Oldenbüttel ; and it eventually crosses the water-parting of the basins of the Eider and the Elbe at Grümenthal, and enters the estuary of the Elbe near Brunsbüttel (Plate 12, Figs. 12 and 13).

**Objects of the Baltic Canal.** When Kiel became the great naval arsenal of Germany, the immense advantage of securing a short, safe, unimpeded passage, through German territory, for vessels of war from Kiel Harbour to the North Sea, instead of having to traverse the Sound or the Belts and go round Cape Skagen, was evident enough. In 1864, directly after the separation of Schleswig-Holstein from Denmark, a project for a canal at one level from Kiel Harbour to the Elbe was prepared; but the wars of 1866 and 1870 necessitated a postponement of the scheme. At length, in 1878, investigations to determine the best route for the proposed canal were recommenced; the construction of the canal was approved in 1886 ; and the works were commenced in 1887.

The construction of the canal was primarily undertaken to strengthen the naval position of Germany in the event of war, by connecting its chief naval arsenal and harbour with the North Sea. At the same time, it has been anticipated that some return on the expenditure will be secured by vessels trading with the German and Russian ports on the Baltic, availing themselves of the canal to shorten their voyage. The Baltic canal is, accordingly, intended to serve a double purpose; its main purpose is similar to the object of the Amsterdam Ship-Canal, though with less peaceful aims; whilst its secondary purpose is similar to that of inter-oceanic or isthmian ship-canals, in shortening the connection between two seas.

**Description of the Baltic Canal.** The canal starts from Holtenau in Kiel Harbour, a large sheltered inlet opening out of Kiel Bay in the Baltic, and following to some, extent the contour of the land, and avoiding severance of property as far as possible, enters the Elbe near Brunsbüttel, after a course of 61⅓ miles (Plate 12, Fig. 12)[1]. The canal has been given a bottom width of 72 feet, with six wider passing places at intervals, side slopes of 3 to 1 near the bottom and 2 to 1 higher up, and a minimum depth of 29½ feet; and a narrow berm below the water-level in deep cutting, and a wide berm

---

[1] A plan, longitudinal section, and cross-sections of the Baltic Canal were sent to me by the engineer-in-chief of the works, Mr. Baensch, and also a description of the works contributed by him to the 'Centralblatt der Bauverwaltung' of February and March, 1889.

in low ground, make the widths of the canal at the water-level 216¾ feet and 283 feet respectively (Plate 11, Fig. 26). The canal has been excavated throughout, except where it traverses two lakes near Rendsburg; and it passes through fairly high ground between Holtenau and these lakes. The deepest cutting, reaching a maximum of 108 feet, traverses the ridge separating the basins of the Eider and the Elbe; but the canal passes through fairly low ground between Rendsburg and the big cutting, and for the last 10 miles in the Elbe basin crosses marshy land barely rising above its water-level, and in places below it. The canal is in one level reach throughout, with regulating locks at both ends; but the bottom, which is level between Holtenau and Rendsburg, is given a fall from thence to the Elbe, gradually increasing from 1 in 200,000 to 1 in 25,000, to allow for the slope of the water surface at ebb tide when the Brunsbüttel lock is left open (Plate 12, Fig. 13). The canal is embanked on both sides where it traverses low-lying land; and the berm at the side has been made wide to allow of the canal being widened without disturbing the embankments. The slope starting from the end of the berms on each side, at 6½ feet below the water-level, is protected by stone up to 3¼ feet above the water-level, where another sloping berm has been formed. The canal is widened out at the curves to an extent ranging from 3¼ feet for curves of 8,200 feet radius, up to 52½ feet for the sharpest curves of 3,300 feet radius[1]. The excavation was effected by twenty excavators and forty-two dredgers, which deposited the dredged material behind the banks by means of long shoots and floating tubes. The total quantity of excavation amounted to about 105,000,000 cubic yards.

The canal is crossed by two fixed, and two opening railway bridges, and one opening roadway bridge near Rendsburg. The fixed bridges have a clear span of 513 feet at Grümenthal,

[1] 'The North and East Sea Canal,' J. Fülscher, Transactions of the American Society of Civil Engineers, vol. xxx, p. 425.

and 536 feet at Levensau, and a clear headway of $137\frac{3}{4}$ feet above the water-level of the canal ; and the hydraulic swing-bridges, one of which is made double for the up and down lines of railway, leave a clear waterway of 164 feet. The Grümenthal bridge is partly an arch and partly a bowstring, as the road-way is suspended from the arch in the central part ; but the Levensau bridge is an arch throughout. The traffic on all the other roads is carried across the canal by fourteen ferries.

**Sand Dams along the Baltic Canal.** Owing to the softness of the soil along the marshy ground, the slopes of the ex-cavated canal would have been very liable to slip, and would have been forced forward by the weight of the embankments on each side, which had to be made to exclude floods from the canal. Accordingly, before excavating the canal through the marshes, sand was excavated from the nearest available cutting beyond the marshes, and tipped by wagons on the bog, into which it sank, forcing aside the soft material, and forming a dam for keeping off the soft stratum at the back from the canal excavations[1]. Where the top of the bog was tough, it was found advisable to remove it, instead of merely loosening it by trenches along the sides, as this upper mattress was liable to be separated into two parts by the tipping in of the sand, which turning up on edge prevented the spreading of the sand. Where, however, the surface was soft, piles were driven along the centre of the dam down to the firm subsoil, which bore the wagons of sand and enabled the sand dam to be carried forward. The sand dams were specially loaded on the edge towards the canal, to reach the maximum depth practicable on this side ; and by tipping in more sand at the sides, the sand dams were finally extended to their full width. The sand dams on each side of the canal were carried across Kuden Lake, 5 miles from the Elbe, by means of rafts, in ad-vance of the completed sand dam, which supported the line on

[1] 'The North and East Sea Canal,' J. Fülscher, *Transactions of the American Society of Civil Engineers*, vol. xxx, pp. 434 to 438, and plates 2 to 5.

which buckets carried on small trucks travelled, conveying the material from the wagons at the extremity of the sand dam, and tipping it on each side of the platform between the end of the dam and the raft, for extending the dam.

These sand dams had to be deposited from $4\frac{3}{4}$ miles from the Elbe end of the canal up to $11\frac{1}{2}$ miles; for though the ground was firmer up to $8\frac{1}{8}$ miles than beyond, and sand was more difficult to obtain, and an attempt was made to form the embankments with dry clay, it was found eventually that the sand dams could not be dispensed with on account of the slipping of the slopes. The sand dams were generally left to settle for six months before the excavations for the canal were commenced. The slopes of the canal near the water-level have been almost wholly excavated in the sand dams, increasing their stability; whilst the embankments have been partially formed on the sand dams, and extend somewhat behind them.

**Locks on the Baltic Canal.** The tides in the Elbe, with a maximum range of $27\frac{1}{2}$ feet between the highest tide and the lowest stage of the river, rendered a lock absolutely necessary at Brunsbüttel; and as the normal water-level in the canal, which is the average level of the Baltic, is $17\frac{1}{8}$ feet below the highest level, and $1c\frac{1}{8}$ feet above the lowest, and about half-way between mean-tide level and ordinary low water in the Elbe at Brunsbüttel, vessels have to be locked both up and down, into and out of the canal (Plate 12, Fig. 13). Two locks have been constructed side by side, one for incoming and the other for outgoing vessels; and each lock is provided with a double pair of steel gates, pointing both ways, at each end and in the centre of the chamber. The lock-chamber, in each case, is 492 feet long and 82 feet wide; and the sills are placed $33\frac{2}{3}$ feet below the normal water-level of the canal (Plate 12, Fig. 14). The quays of the locks have been raised 5 feet above the highest level of the Elbe, to secure the low-lying lands at the back along the canal against floods; and longitudinal culverts have been constructed in the side walls of the locks, with several branch culverts opening into the lock-chamber, to enable the locking of vessels to be effected rapidly. The canal is widened out at the inner end of the locks for the accommodation of vessels; and the wide outlet channel beyond the locks is guided by jetties, on each side, into the Elbe.

Two precisely similar locks have been constructed at Holtenau, opening into Kiel Harbour; for though generally the variations in the level of the Baltic are small, the sea may rise in storms $10\frac{2}{5}$ feet above the mean level, or fall $6\frac{4}{5}$ feet below it (Plate 12, Figs. 13 and 15). The lock-gates, accordingly, are usually kept open at the Baltic end; and locking is only resorted to when the level of the Baltic exceeds $1\frac{2}{3}$ feet above or below its mean level. Moreover, as the Baltic never rises within $6\frac{1}{2}$ feet of the highest level in the Elbe, as the locks at Holtenau are much more sheltered than at Brunsbüttel, and as there is no low-lying land to be protected from floods near the Baltic end of the canal, the quay level of the Holtenau locks has been placed 9 feet lower than that of the Brunsbüttel locks.

The Lower Eider was formerly in communication with the Baltic by the Eider Canal, which the Baltic Canal has absorbed; but though the lowering of the Eider Lakes above Rendsburg, and of the Lower Eider at Rendsburg, to the extent of 7 feet in the construction of the canal, has brought the average water-level of the Lower Eider down to the water-level of the canal, the variations in the level of the Lower Eider at Rendsburg, caused by tides and wind, has necessitated the construction of a lock near Rendsburg where the Lower Eider is connected with the canal. This lock has an available length of 223 feet, a width of $39\frac{1}{3}$ feet, and a depth of water on the sills of $17\frac{1}{4}$ feet; and it is provided with a double pair of gates pointing both ways at the lower end, and a pair of fan gates of a type first used in Holland, at the upper end, to enable vessels to be locked up or down, according to the relative levels of the canal and the river. Each fan gate consists of two gates rigidly connected at an angle of .70°, one the actual gate, and the other a wing with a rather greater width, revolving on the heel-post into and out of a recess in the side wall formed like a quadrant [1]. By causing water, admitted through sluices in the lock wall, to press upon one or the other side of the wing, the wing is turned to the front or back of the recess, closing or opening the gate even against a current of water. These fan gates, accordingly, can be closed when a current is running through the locks with the lower gates open, and also serve for locking in either direction, up or down.

**Levels of Lakes lowered for Baltic Canal.** The three lakes above Rendsburg had to be lowered 7 feet, in order to conform with the water-level fixed for the Baltic Canal which passes through them.

The canal passes near Rosenkranz, close to the Flemhude Lake, 580 acres in area and the summit-level of the Eider Canal, whose water-level was 23 feet above the water-level proposed for the canal. The maintenance of this difference of level close to the canal was inadmissible ; but, at the same time, the lands bordering the lake could not be deprived of their supply of water. A dam, accordingly, was formed skirting the shore of the lake, enclosing an encircling canal along the margin, with its water-level kept up at the original level of the lake, being replenished by the Eider which flows into the encircling canal as it formerly did into the lake; and the surplus water of the river falls through a cut into the lake, whose level has been lowered 23 feet, which lowering, and also the deposit of material from the excavations for the canal, have reduced the area of the lake to one-third of its former extent, thereby reclaiming land.

The Meckel Lake, near Breiholz, through the centre of

[1] 'Engineering,' vol. lx, p. 35.

which the canal passes, has similarly been provided with a canal round its margin, fed by the Haarbe, a tributary of the Eider; and the remainder of the lake, with the exception of the canal widened for a passing place, has been reclaimed by the deposit of material excavated from the canal.

A portion of the Kuden Lake in the marshes, through which the canal passes, has been reclaimed in the neighbourhood of the canal by the deposit of dredgings from the canal.

**Completion and Cost of the Baltic Canal.** Preliminary preparations for the construction of the canal were commenced towards the end of 1886, the land was acquired in 1887, the earthwork was commenced early in 1888, the locks in the autumn of 1890, and the bridges in the spring of 1891 ; and the canal was formally opened in June, 1895. The canal is lighted at night by electric lamps, placed at intervals of about 800 feet, and closer round the curves; and the channel through the lakes is indicated by light-giving buoys. The time occupied in passing through the canal has been fixed at 10 to 12 hours, to avoid too great a wash on the banks ; but in an emergency, the canal could be traversed much more rapidly.

The works are remarkable for having been completed by the time originally specified, and with an expenditure slightly within the estimated cost of £8,000,000.

**Remarks on the Baltic Canal.** The canal, by cutting across Holstein, will shorten the route of all vessels trading between the Baltic ports and all ports the direction of whose approach lies south of Newcastle, which have hitherto had to go round through the Sound and the Skager Rack. The saving in distance by the canal is naturally greatest for Hamburg and Bremen, amounting to 425 and 323 nautical miles, or 45 and 32 hours respectively; whereas the saving to Amsterdam, Rotterdam, Antwerp, Dunkirk, London, and practically all ports reached through the English Channel, amounts to about 238 nautical miles, or 22 hours, and even to Newcastle and Leith, 107 and 84 miles, or 6 and 3 hours respectively. Assuming, however, that all vessels engaged in the Baltic trade running to ports to the south of Hartlepool, should use the canal, the tonnage of vessels passing through the canal would amount to two-thirds of the whole tonnage of this trade. On the other hand, it has been stated that British shipowners consider that the comparison of actual distances leads to erroneous conclusions, and that a good deal of time would probably be lost in navigating the shallow, narrow, winding channels of the Elbe, as well as in passing through the canal, so that the actual saving in time by the canal is uncertain[1]. Moreover, many British ships call at Copenhagen on their way to and from the Baltic, 50 miles to the north of the common starting-point of the distances at the island of Moën in the Baltic, a course which if combined with the passage through the canal, would greatly reduce the saving of distance and time. The dangers also of rounding Cape Skagen, which have been urged as greatly enhancing the value of the canal, are said to have been very much reduced by buoying, lighting, and good pilotage. Time alone will show to what extent British trade will use the canal; and the traffic on the canal must eventually greatly depend upon the actual saving of time on the one hand, as compared with the amount of the dues on the other.

In any case, Germany has obtained a most valuable route for her navy in the event of hostilities breaking out; and the advantages gained by the use of this route might perhaps recoup the expenditure in a single war. The canal is also very valuable in greatly facilitating the communication between the German ports on the Baltic and on the North Sea.

**Comparison of the Baltic and Amsterdam Ship-Canals.** The Baltic and Amsterdam ship-canals both lead from an inland sea to the North Sea, both pass for some distance through low ground or lakes, both have one level reach from end to

---

[1] 'The Times,' 'British Shipping and the Baltic Canal,' July 10, 1895.

end, both have necessitated a maximum depth of cutting of about 100 feet, and both have regulating locks at each end (Plate 12, Figs. 1 and 12). The canals have about the same width at the water-level, and wide berms at the side in low ground (Plate 11, Figs. 25 and 26); but the enlarged Amsterdam Ship-Canal has the greatest bottom width, though slightly less depth. The Baltic Canal is considerably the longer and larger work of the two, though involving fewer auxiliary works than the other; and it is, at the same time, the more important work, as having larger capabilities of trade, with the numerous ports of the Baltic behind it. The Amsterdam Ship-Canal was constructed, indeed, by a Company, whose financial difficulties led to its being handed over to the State, and has been freed of tolls to foster the trade of Amsterdam; whilst the Baltic Canal had to be carried out by the Government, as a Company could not be formed for its construction; and it is hoped that the tolls on the traffic passing through it will provide a fair return on the expenditure. The Baltic Canal has the great advantage of serving the double purpose of facilitating naval operations, and shortening the commercial routes over an extensive district; whereas the Amsterdam Ship-Canal only accommodates Amsterdam, which, however, would have ceased to be a port without the deep waterway it provides. The Amsterdam Ship-Canal was more costly than the Baltic Canal in proportion to its length, and occupied a longer time in construction; but these results may be attributed to the large auxiliary works required for the Amsterdam Ship-Canal, comprising the embankment, the pumping machinery, and the North Sea harbour, and the difficulties experienced by the Company in raising the necessary funds.

### MANCHESTER SHIP-CANAL.

Manchester with a larger population than Liverpool, with extensive cotton manufactories, and the centre of a manufacturing district, naturally wished to import the raw materials and export its manufactures without having to pay the high Liverpool dock dues and heavy railway rates. The Mersey, however, has a very shallow, shifting channel though its upper estuary above Garston and Eastham at low tide, and is very narrow, shallow, and tortuous above Warrington. Communication by water between Manchester and the sea was really confined to the Bridgewater Canal entering the Mersey at Runcorn, and the Leeds and Liverpool Canal joining the Mersey at Liverpool; but these canals, though carrying a large traffic, are only inland canals of very moderate dimensions, involving transhipment for sea-going trade.

Schemes for Ship-Canal. Early in the nineteenth century a proposal was made to form a canal from Manchester, across Cheshire, to the estuary of the Dee; and in 1877, the Manchester Chamber of Commerce urged the necessity of making a ship-canal from Manchester to the sea. No actual steps, however, were taken till 1882, when two schemes were submitted. One scheme was designed to deepen and straighten the Mersey and the Irwell from Liverpool to Trafford Bridge at Manchester, guiding the channel through the upper estuary of the Mersey by training walls, so as to provide a channel 22 feet deep at low water of spring tides, and bring the tide up to Manchester, thereby dispensing with locks, but necessitating a very deep cutting for the proposed docks at Manchester, as the ground rises about 60 feet between Liverpool and Manchester. The second scheme proposed the deepening of the Mersey estuary from deep water near Garston up to Runcorn by training walls and dredging, and the deepening and straightening of the river from Runcorn up to Latchford, from whence a canal was to be formed, 22 feet deep and 100 feet wide at the bottom, along the general course of the Mersey and Irwell in a fairly direct route, ascending by three locks to Manchester, the estimated cost of this scheme being £5,160,000; and this was the scheme which was adopted in principle. In the scheme submitted to Parliament in 1883, the canal was to leave the Mersey 1½ miles above Runcorn ; but this scheme was rejected, as there was no indication on the plans of the position of the proposed training walls below Runcorn. In 1884, a scheme was brought before Parliament showing training walls in the Upper Mersey estuary (Plate 10, Fig. 8) ; a tidal lock was introduced at the entrance to the canal above Runcorn ; the canal was somewhat straightened for some miles above Runcorn ; and the canal was to rise to Manchester by four locks instead of three, resembling the actual canal between Latchford and Manchester. This scheme, however, shared the fate of the first, on account of the danger that the training walls in the upper estuary might lead to extensive accretions at the sides, whilst deepening the channel, and thus produce a raising of the Mersey bar by the diminution of the tidal scour from the upper estuary (p. 340, and Plate 10, Fig. 8).

At last, in 1885, the promoters proposed forming an outlet channel for the canal along the Cheshire shore of the estuary, from above Runcorn to Eastham, and entering the estuary where deep water approaches the shore and leads into the Sloyne Deeps ; and this scheme was authorised, and has been carried out (Plate 12, Figs. 7 and 8).

**Description of the Manchester Ship-Canal.** The canal is entered at Eastham through three tidal locks side by side ; and the tide flows up the first reach of the canal between Eastham and the Latchford locks, 21 miles long, whenever the tide rises over 9½ feet above mean-tide level at Eastham, which is the normal water-level retained in the canal by the lock-gates at Eastham, 26 feet above the bottom of the canal along this reach, which is the depth of water provided throughout the rest of the canal (Plate 12, Fig. 8). The locks at Latchford have a lift of 16½ feet, so that the water-level in the second reach of the canal up to the Irlam locks, 7½ miles long, is 16½ feet higher than the normal water-level in the tidal reach, or 26 feet above mean-tide level at Eastham. The Irlam locks retain the water-level in the third reach 16 feet higher, as far as Barton, a distance of 2 miles, where there is another lift of 15 feet extending 3½ miles to the last locks at Mode Wheel, 1¾ miles from the Manchester docks, with a lift to this reach of 13 feet, making the water-level in the fifth reach, and in the docks at Salford and Manchester, 70 feet above mean-tide level at Eastham[1]. The total length of the canal, accordingly, between Eastham and Manchester, is 35½ miles, with a rise of 60½ feet; and the canal attains a depth of about 33 feet in the tidal reach at the highest spring tides.

The section of the canal varies in several places to provide for special conditions; but the general section, between Eastham and the Barton locks, has widths of 120 feet at the bottom, and 172 feet at the water-level, with slopes of 1 to 1, protected by pitching from the bottom to above the water-line (Plate 11, Fig. 20); and the narrowest part of the canal is under Runcorn Bridge, between the first river pier of the bridge and the river wall of the canal, where the width is reduced to 92 feet. The canal is widened out between the Barton locks and Manchester to a bottom width of 170 feet. As the tidal reach of the canal skirts the estuary, 9 miles of embankments have had to be constructed on the outer side, along the portions of the cutting open to the estuary. The embankments have been made 30 feet wide at the top, with an outer slope of 1½ to 1, protected by pitching, and an inner slope of 1 to 1; and where the foundation is sand, sheet piling has been driven along the toe of the embankment on each side, to prevent percolation and undermining. At Runcorn, where the width is limited, a concrete wall has been substituted for the embankment, which has been made 22 feet wide at the base.

---

[1] A plan, longitudinal section, and cross-section of the canal were given me by Sir E. Leader Williams, from which Figs. 7 and 8 of Plate 12, and Fig. 20 of Plate 11, have been reduced.

Water is supplied to the upper reaches of the canal by the Mersey and Irwell, which are to a great extent absorbed into the canal from about three miles above Latchford up to Manchester; but the admission of the polluted waters of the Irwell and of the Mersey into the canal has caused deposit, and has proved otherwise objectionable, rendering the purification of these rivers very urgent.

The available headway under fixed bridges for vessels navigating the canal was determined by Runcorn Bridge, which leaves a clear height of only 75 feet between the normal water-level in the tidal reach and the underside of the girders; and, therefore, the other fixed bridges have been placed at the same height.

**Excavation of the Manchester Ship-Canal.** The excavation of the canal was mainly effected by steam navvies, steam-crane navvies, and French and German bucket-ladder excavators (pages 95 to 98). Grabs were used principally for excavating foundations; and bucket-ladder dredgers were employed for excavating in front of Ellesmere Port where the waterway had to be kept open till the canal below was ready to take the traffic, for forming an outlet channel below Eastham locks, and in deepening the canal after it was flooded in 1890 (Plate 2, Fig. 1). Altogether, in excavating the 53,500,000 cubic yards for forming the canal, ninety-seven excavators, eight large bucket-ladder dredgers, and some smaller dredgers and grabs were employed, the main portion of the work having been accomplished by fifty-eight steam navvies which could excavate soft sandstone rock. A soil transporter, consisting of a long endless travelling band, was used in some parts for conveying the dredged material from the dredger in the canal to wagons on the bank. Of the 41,500,000 cubic yards of soft soil removed, 38,500,000 cubic yards were excavated and 3,000,000 cubic yards dredged; and of the 12,000,000 cubic yards of sandstone rock removed, the harder portions of which had to be blasted, 11,500,000 cubic yards were excavated and 500,000 cubic yards were dredged[1].

**Locks and Sluices on the Manchester Ship-Canal.** At Eastham, there are three locks of different dimensions side by side, and an outer pair of storm gates at each lock pointing outwards to protect the locks from waves coming up the estuary; but at the four other sets of locks, there are only two locks side by side. The locks at Eastham have chambers 600 feet long by 80 feet wide, 350 feet by 50 feet, and 150 feet by 30 feet, with sills 28 feet, 25 feet, and 16 feet respectively, below the normal water-level of the canal (Plate 12, Fig. 9). The approach channel to the locks has been dredged down to 33 feet below the highest tide, and 23½ feet below high water of neap tides, or about the level of the bottom of the canal above the locks; but the outer sills of the locks have been laid 11 feet lower to allow for further deepening of the channel[2]. Vessels, accordingly, of large draught can enter or leave the canal from over an hour before till more than an hour after high water during spring tides ; and different sizes of vessels are provided for by the variety in the dimensions of the locks. Two sluices, 20 feet wide, have been constructed between the large lock at Eastham and the land, for assisting the influx of the tidal water above the normal level of the canal, which also flows in through the locks and the tidal openings in the embankments. These sluices, and the other sluices along the canal are closed and opened by counterbalanced sluice-gates sliding on free rollers (p. 115, and Plate 3, Figs. 17 and 18). The sills of the sluices at Eastham are level with the bottom of the canal ; and the sluice-gates can be raised 15 feet.

The double locks at Latchford, Irlam, Barton, and Mode Wheel, are all of the same size, with chambers 600 feet by

---

[1] 'The Manchester Ship-Canal,' reprinted from 'Engineering,' January 26, 1894, p. 30.

[2] 'The Manchester Ship-Canal,' E. Leader Williams, Manchester Inland Navigation Congress, 1890.

65 feet, and 350 feet by 45 feet, but with rises varying from 16½ feet at Latchford to 13 feet at Mode Wheel; and they all have an intermediate pair of gates to expedite the locking of small vessels (Plate 12, Fig. 10 and 11). The sills of all these locks, as well as the inner sill of the large Eastham lock, have been placed 28 feet below the ordinary water-level of the canal, to admit of a future deepening of the canal from 26 to 28 feet. The gates of all the locks have been constructed of greenheart, and are worked by hydraulic machinery. As the floods of the Mersey and Irwell come down the upper portion of the canal, provision has been made for their discharge by sluices alongside the locks, each 30 feet wide, closed by sluice-gates sliding upon free rollers, and capable of being raised about 20 feet. There are only three of these sluices at Latchford (Plate 12, Fig. 11), as the flood water is discharged into the river where it leaves the canal about 2 miles above Latchford. At Irlam, five sluices have been provided for dealing with the floods, as the flow of both the Mersey and Irwell passes this lock; and four sluices have been built alongside the Barton and Mode Wheel locks, as the Irwell alone supplies the canal at this part. The canal has been widened out near the locks to facilitate the passage or turning round of vessels, and to accommodate vessels waiting to enter the locks.

**Sluices and Tidal Openings into the Mersey Estuary.** The river Weaver has been shut off from the Mersey estuary by an embankment, for retaining the water in the canal, passing in front of the mouth of the Weaver, so that the former tidal reach of the river up to Frodsham, 3 miles from the canal, is kept up to the normal water-level of the canal, where the traffic of the Weaver Navigation enters the canal through a lock, 229 feet by 42 feet, enabling the salt trade of Cheshire to get down to the canal. In order that the Mersey estuary may not suffer from the abstraction of the tidal flow which formerly passed up and down the Weaver at this part, sluices had to be erected to discharge the tidal water, flowing from the river into the canal during the ebb of tides rising above the normal water-level, into the estuary at the same period as formerly. Accordingly, ten sluices, each 30 feet wide, have been constructed in a row in the embankment between the canal and the estuary in face of the Weaver, closed by counterbalanced sluice-gates, 16 feet high, sliding on free rollers. The sills of these sluices are placed 16½ feet above the bottom of the canal ; and the sluice-gates are raised after high water at every high tide to an extent corresponding to the required discharge with the particular rise of tide, having a maximum lift of 18 feet.

Two similar sluices, each 30 feet wide, have been provided at Old Randles, about 3 miles above Runcorn, where the canal comes close to the river, to let water out of the canal, whose sills are 6⅛ feet below the normal water-level of the canal; and the lift of the sluice-gates is 10 feet. These sluices, together with the Weaver sluices, enable the efflux of the surplus tidal water in the canal at high tides to be regulated, and to be discharged, if desired, into the estuary at the most favourable period for scouring the channels.

Three tidal openings, each 600 feet wide, formed in the embankments between the canal and the estuary, with their sills level with the normal water-level in the canal, provide, in addition to the sluices, for the free influx and efflux of the water along the tidal reach of the canal, for all tides rising higher than 14⅛ feet above Old Dock Sill in the upper estuary. These tidal openings are situated opposite Ellesmere Port, just below the Weaver sluices, and a little over half a mile above Runcorn Bridge. The tide, however, entering the canal through Eastham lock and the lower tidal openings, flows quicker up the deep channel of the canal than up the shallow estuary encumbered with sandbanks, so that the water in the canal at high tides begins to flow over the upper tidal opening before the tide in the estuary has risen to the sill of the opening. Moreover, the water entering the upper tidal opening washes in sand from the adjacent high sandbanks, which deposits in the canal ; and the current also through the openings at the highest tides is somewhat inconvenient for navigation during a short period. As the sluices at the Weaver, Old Randles, and Walton Lock could easily deal with the discharge which at present takes place through the tidal openings, and let it into the estuary at a more favourable period, it appears that the tidal openings at Eastham, the Weaver, and above Runcorn might be closed without detriment to the estuary. The closing also of the opening above Runcorn would secure the estuary above from the diversion into the canal, through this opening, of some of the limited tidal flow passing through Runcorn Gap.

**Bridges across the Manchester Ship-Canal.** As the canal skirts the estuary nearly up to Warrington, the railway bridge at Runcorn Gap is the first bridge that crosses the canal, at a distance of 12½ miles above Eastham. The first bridge above Eastham erected by the Canal Company across the canal is a swing-bridge, about half a mile above the railway bridge, leading to the Old Quay at Runcorn ; and no other bridge had to be constructed before reaching the Moore Lane swing-bridge, 17 miles above Eastham. Including this latter bridge, twelve bridges and a swing aqueduct had to be erected in the last 18½ miles up to Manchester, for crossing the canal.

All the four bridges carrying the railways over the canal have been made high-level bridges, with a clear headway of 75 feet above the water-level of the canal, as swing-bridges would have impeded the very heavy traffic along these lines ; and two of the roadway bridges have been also made high-level bridges, namely, one near Latchford, and the other at Warburton, between Latchford and Irlam locks. The remaining seven main roads cross the canal on swing-bridges. As the surface of the ground nowhere rises any considerable height above the canal, the high-level bridges have involved long, costly approaches to reach the height imposed by the required headway of 75 feet. Moreover, as the traffic on the railways crossing the line of the canal could not be interfered with, the railway bridges in each case were built along a deviation of the railway, on to which the traffic was permanently diverted after the completion of the bridge, before the original lines could be touched for the formation of the canal across them. Ferries serve for conveying the traffic of minor roads, and pedestrians using old rights of way, across the canal.

A swing aqueduct carries the Bridgewater Canal over the ship-canal, close to the place where the Barton masonry aqueduct crossed the valley of the Irwell, as the Bridgewater Canal is at too low a level to provide the requisite headway for the ship-canal traffic. The Barton swing aqueduct, like the Barton roadway swing-bridge close by, turns on a central pier in the middle of the canal which has been widened to allow vessels to pass on each side[1]. The swing aqueduct consists of an iron trough, 19 feet wide and 7 feet deep, with 6 feet of water in it, and a towing-path at the side, carried by two main girders, 234½ feet long and 22¼ feet apart, braced overhead. The trough is closed at each end by iron gates with a watertight joint when it has to be swung, and similar gates close the canal at each end; and watertight joints are formed between the ends of the trough and the canal when the trough is in line with the canal and the gates opened. The swinging portion of the aqueduct, with its load of water, weighs about 1,600 tons; and it is turned, like the swing-bridges and lock-gates, by hydraulic machinery. The swing aqueduct, in fact, resembles in principle the hydraulic canal lifts previously described, except that it is swung round instead of being lifted vertically.

**Various Works of the Manchester Ship-Canal.** There are several works which, though not strictly forming part of the canal itself, were necessitated by its construction, or are

---

[1] 'The Manchester Ship-Canal,' reprinted from 'Engineering,' January 26, 1894, p. 26, and plate 2.

supplementary to it. Thus where the tidal reach of the canal crosses streams which discharge into the Mersey estuary, it was necessary to carry these streams under the canal in an inverted siphon, owing to the water-level of the canal being always kept up to 14⅙ feet above Old Dock Sill. Accordingly, the river Gowy, the largest of these streams, passes under the canal in two cast-iron siphons 12 feet in diameter; and similar provision in iron or masonry has been made for the discharge of Pool Hall Brook, and other minor streams.

A lock has been constructed at Weston Marsh, connecting the Weston Canal with the ship-canal, to which it runs parallel from near the mouth of the Weaver to the Weston Docks at Weston Point. At Weston Point, there is a lock from the ship-canal into the Mersey, to enable vessels to pass straight from the Weston Docks into the Mersey. This Weston Mersey lock has a chamber 600 feet long and 45 feet wide, provided with intermediate gates. Another lock, 450 feet long and 45 feet wide, leads into the Mersey opposite the Bridgewater Docks, 11¾ miles above Eastham; and this Bridgewater lock connects the Bridgewater Canal with the Mersey across the ship-canal. The Runcorn lock, 250 feet long and 45 feet wide, leads out of the ship-canal about half a mile above Runcorn Bridge, and affords direct access to the Mersey from the wharves at Runcorn; and this lock also enables barges from the Widnes alkali works to enter the ship-canal. A lock has been erected at the lower end of a bend of the Mersey, connected with the canal, about 19 miles above Eastham, for which a straight cut has been substituted. This Walton Lock, with sluices alongside, keeps up the water in this old reach which has been cut off from the river, and which was intended originally to be formed into a dock for Warrington; the lock provides a communication between the canal and the river; and the sluices furnish a means of scouring the river below the lock (Plate 12, Fig. 7). A small lock, 85 feet long and 19½ feet wide, with a rise of 12¾ feet, connects the Latchford

Canal with the ship-canal about half a mile above the Walton lock approach.

The Manchester and Salford Docks, forming branch basins opening out of the ship-canal between Mode Wheel locks and Manchester, have a total area of 104½ acres, 5¼ miles of quays, and 152 acres of quay space.

Cost of the Manchester Ship-Canal. After very heavy parliamentary contests in three successive sessions, the Act for the construction of the ship-canal was obtained in 1885; but there was great difficulty in raising the necessary capital, one of the conditions imposed upon the Canal Company being that £5,000,000 must be raised in shares before the work could be commenced. At last, in 1887, this necessary condition was fulfilled; and the works were let for a contract sum of £5,750,000 in June, 1887; the Bridgewater Canal was purchased, in accordance with the terms of the Act, for £1,710,000 soon after; the works were commenced in November, 1887; and the canal was opened for traffic, right up to Manchester, on January 1, 1894, though it was opened up to Ellesmere Port in July, 1891, and a year later up to Saltport, close to the Weaver. The total cost of the works, including land, the Bridgewater Canal, parliamentary and other expenses, and interest on capital during construction, has reached the large amount of about £15,000,000. The original capital amounted to about £10,412,000 with debentures, to which has been added a loan from the Corporation of £5,000,000, the cost of the actual works, with equipment and maintenance, having risen to over £10,000,000 from a variety of causes. The death of the contractor when the easier portion of the work only had been done, which, by the eventual termination of the contract, threw the more difficult portions on to the Company, the floods of 1890, additional works found necessary, slips in the cuttings, the less favourable nature of the sandstone excavated than anticipated for serving as pitching, and an unexpected amount of dredging for maintenance, together with a delay of two years in the completion of the canal, entailing additional expenditure in interest during construction, all combined to swell the cost. The yearly cost at present of dredging alone for maintenance amounts to from £40,000 to £50,000; but when the Mersey and Irwell are purified, it is probable that much less dredging will be necessary in the upper part of the canal.

Traffic on the Manchester Ship-Canal. The canal has not been opened for a sufficient time to enable an estimate to be formed of its probable future traffic. The weight of merchandise carried by sea-going vessels over the canal had nearly doubled in the first half of 1895, as compared with the corresponding period in 1894, having risen from 261,100 tons in the earlier half of 1894 to 480,800 tons in 1895; and the weights carried in barges also increased considerably in the same period, the totals for the first six months of 1894 and 1895 being 369,967 tons and 600,100 tons respectively. A very much larger traffic and net revenue, however, will be required to put the canal on at all a sound financial basis.

Remarks on the Manchester Ship-Canal. The Manchester Ship-Canal is the first large ship-canal which has been constructed with locks, raising the vessels 60½ feet, and transporting them inland, and thereby converting an inland city into a seaport. The other large ship-canals have either no locks at all, as in the case of the Suez and Corinth canals, or only regulating locks at each end to provide for small changes of level in the sea due to tide or wind, as at the Amsterdam and Baltic canals; though locks were eventually proposed for the Panama Canal to enable it to surmount the Culebra ridge with less excavation, and locks form an essential part in the design for the Nicaragua Canal.

The Manchester Ship-Canal has had the disadvantage of having to traverse land abounding in vested interests, necessitating a very large expenditure in complying with the requirements of the bodies possessing these rights. The cost of auxiliary and compensating works were difficult to estimate;

and the magnitude of the works rendered it difficult to raise the capital on favourable terms, and made it impossible to foresee all the contingencies that might arise in the course of construction; whilst improvements and modifications in the course of execution aided in increasing the cost. Unfortunately, hardly any very large undertakings resemble the Baltic Ship-Canal in being carried out within the estimated cost; and this work possessed the advantage of being a national undertaking, for which provision was made without difficulty out of public funds, as well as traversing a much less populous district than the Manchester Ship-Canal. The experience, indeed, of Manchester and Preston will check the ambition of authorities of towns some miles from the sea, desiring to convert their towns into seaports; but such undertakings must necessarily be embarked upon with great caution, as it is very difficult to estimate their cost correctly, and very hard to foresee to what extent shipping and trade will avail themselves of the facilities offered. The development of a new route for shipping is necessarily slow; and though the Manchester Ship-Canal cannot expect, from its position, to rival the final success of the Suez Canal, which in its early stages had its financial difficulties, it is probable that, with the strenuous aid of the merchants of Manchester and increased facilities for shipping, the canal will eventually prove an important assistance to the commercial prosperity of Manchester. In any case, the energy of Manchester is worthy of commendation in having, in the face of so many difficulties and discouragements, brought the ship-canal to a successful completion. In spite of the unfavourable financial aspect of the case, the works themselves must be regarded as a remarkable engineering achievement.

## BRUGES SHIP-CANAL.

Works for the conversion of Bruges into a seaport by the construction of a ship-canal to the North Sea at Heyst, with a sheltered approach on the sea-coast, have been recently commenced.

**Object of the Bruges Ship-Canal.**　By giving Bruges deep-water access to the sea, it is hoped that the commercial importance which Bruges possessed in the middle ages may be revived, and that it may again become a flourishing centre of trade. The canal may also possibly be regarded as the first step towards providing Ghent with a more direct water-way to the sea, without passing through Dutch territory.

**Description of the Bruges Ship-Canal.**　A breakwater starting from the coast, to the west of the entrance of the canal at Heyst, in a northerly direction, will curve round to an east-north-easterly course; and extending out opposite the entrance to the canal, and being provided with quays, it will both protect the approach from the exposed quarters of west round to north, and also provide a sheltered harbour for the accommodation of ocean-going vessels. The canal, traversing flat low-lying land throughout, will proceed in a direct southerly line from the sea-coast to Bruges, a distance of 7½ miles, where it will branch out into two basins just outside the town [1]　The canal is designed to have a width of 72 feet at the bottom, and 131¼ feet at the water-level, a depth of 26¼ feet, and side slopes of 3 to 1, pitched with stone a little above and below the surface of the water. A lock to provide for the tidal variations in the sea-level is to be constructed a short distance inland of the entrance, with a width of 65¾ feet between the vertical walls at the two ends, a chamber 518 feet long with a bottom width of 79 feet, and an available length of 840 feet between the caisson gates at each end.

The estimated cost of the work is £1,560,000, of which £1,080,000 will be provided by the State for the harbour at Heyst, and the remainder by Bruges, and by a company which will work the canal and receive the dues for loading

---

[1] Plans of the canal and harbour as designed were sent to me by Mr. A. Dufourny, Ingénieur principal des Ponts et Chaussées at Brussels.

and unloading vessels in the harbour and docks for a period of 75 years. The passage along the canal is to be free of toll.

**Remarks.** Heyst was chosen for the outlet of the canal as depths of 23 feet at low tide approach nearer the coast there than at Ostend. The prospects of traffic on this new waterway are not favourable, as Bruges possesses little trade, and there is very little traffic on the inland canals connecting it with ports on the sea-coast. Moreover, Ghent has secured a good maritime trade along the Ghent-Terneuzen Canal, and Antwerp has become one of the chief ports of continental Europe; whilst Dunkirk, on the other side of Heyst, has attracted a considerable trade. If Bruges had obtained a deep-water access to the sea several years ago, it might possibly have procured some of this trade; but it does not possess any specially favourable conditions likely to enable it to rival the well-situated flourishing ports in its neighbourhood. An extension, however, of the ship-canal to Ghent would greatly improve the prospects of traffic along this waterway.

# CHAPTER XXIV.

## INTEROCEANIC SHIP-CANALS.

ISTHMIAN ship-canals constitute some of the most remarkable enterprises of the nineteenth century; but, except as regards the size and depth of their channels, they cannot be regarded as entirely novel conceptions. Herodotus, indeed, records a proposal for cutting through the Isthmus of Suez; a means of communication by water for small vessels appears to have been formed about 600 B.C., and maintained for some centuries; and traces have been found of an ancient canal in places near the line of the Suez Canal, between El Guisr and the Red Sea. The idea of cutting a channel for vessels across the Isthmus of Corinth originated in very early times, when a direct communication between the Ionian Sea and the Archipelago would have been of the highest importance; a canal was actually commenced in the reign of Nero, but his death put a stop to the works; and the Greeks foreshadowed the system of ship-railways, by dragging their triremes across the isthmus over polished granite ways. The proposal also to connect the Atlantic and Pacific oceans, by a cut across the Isthmus of Panama, appears to have been started very soon after the discovery of America, owing to the obvious advantages of such a means of communication.

**Canal des Deux Mers.** The first canal carried out for connecting two seas, which is still in existence, is the Languedoc Canal, or Canal du Midi (p. 481), which, by providing a direct waterway across France between the Bay of Biscay and the Mediterranean, was designed to enable sea-going vessels to effect a saving in distance of about 750 miles, by avoiding the detour through the Straits of Gibraltar. This canal, however, though regarded as a remarkable undertaking in the seventeenth century, only possesses the dimensions now regarded as necessary for the main lines of inland navigation in France. The chief interest of the work at the present day, therefore, lies in its being the first of this class of undertakings carried out by the French, a nation which has since taken the foremost position in designing and carrying out isthmian canals, and also in the proposals made within recent years to reconstruct this waterway, so as to make it accessible to the largest ocean-going vessels. An ample, unrestricted waterway across the south of France would afford similar strategical and commercial advantages to France as the Baltic Canal does to Germany. The physical conditions, however, of the two sites are very different. As the ground rises over 600 feet above the sea at Naurouse, several locks would be needed; the length of waterway to be formed or improved would amount to about 300 miles; and there are

numerous roads and railways traversing the line of the canal,
which would have to be accommodated with bridges. The
slow speed at which vessels could traverse such a canal,
together with the delays at the locks, would neutralize the
saving in distance for vessels of high speed; and the
difficulties and vast cost of the work render its execution
impracticable under reasonable conditions [1].

**Caledonian Canal.** This canal, providing communication
between the North Sea and the Atlantic Ocean across the
Highlands of Scotland, was constructed by the Government
in 1804–1823, in order to enable vessels of war to pass
rapidly from one sea to the other, to afford merchant vessels
a place of refuge from privateers in time of war, and to save
sailing vessels the delays and danger of making the circuit
round the north of Scotland through the stormy Pentland
Firth. The canal extends from Clachnacharry, near Inverness,
on an inlet from Inverness Firth on the east coast, to Corpach,
near Fort William, on Loch Eil which is in connection with
the Atlantic on the west coast, traversing a low, straight
valley stretching from north-east to south-west and containing
a remarkable chain of long and comparatively narrow lakes [2].
By utilizing these lakes, which together extend over 38½
miles of the total length of the waterway of 60½ miles, only
22 miles of canal had to be constructed for completing the
communication between the lakes and the firths at both ends.
Loch Oich forms the summit-level of the canal, with its
water-level 94 feet above high water at either end; and it
is situated between Loch Ness, with a length of 22 miles
and a maximum depth of 774 feet, and Loch Lochy, with
a maximum depth of 456 feet and 10 miles long.

There is a tidal lock at each extremity of the canal, and
there are also three regulating locks at the entrances to the
lakes; and the differences in level are surmounted by eleven

[1] 'Revue des Deux Mondes' Nov. 15, 1893, pp. 338–348.
[2] 'Life of T. Telford, C. E., written by himself,' 1838, pp. 49 and 297.

locks on the North Sea side of the summit-level, and twelve
locks on the Atlantic side, the most remarkable of these being
a flight of eight locks at Banavie, near Loch Eil. The locks
are 160 to 170 feet long, 38 to 40 feet wide, and 17 feet depth
of water on the sills; and the rises of the locks are mostly
between 7½ and 8½ feet [1].

The canal was designed to be made 50 feet wide at the
bottom and 120 feet at the water-level, and 20 feet deep, so
as to give access to vessels of 1000 tons (Plate 11, Fig. 19).
In order, however, to complete the canal without further delay,
and to reduce the cost, the canal was not excavated to its
full depth in the cutting at the summit-level and in the
shallow parts of Loch Oich, so that it can only be traversed
by vessels not exceeding 160 feet in length and 38 feet in
width, and having a maximum draught of 15 feet.

The original cost of the works was about £1,000,000; but
the total expenditure on the canal by the Government, up to
1889, amounted to £1,254,047. The traffic on the canal was
never large; it now amounts to only about 166,000 tons
annually; and it is mainly composed of steamers conveying
tourists in the summer-time to view the scenery. The maintenance and working expenses of the canal exceed the receipts;
and the deficit appears to be made up by grants from the
Treasury. Like the North Holland Canal, the section of the
Caledonian Canal shows, by a comparison with sections of
the ship-canals of the present day, how greatly the dimensions
of sea-going vessels have increased within the last eighty
years (Plate 11, compare Figs. 19 and 24 with Figs. 20, 25,
26, and 32); and consequently ship-canals of an earlier
period are quite inadequate for the requirements of the large
steamers which now trade with distant ports.

---

[1] 'Returns made to the Board of Trade in respect of the Canals and Navigations
in the United Kingdom for the year 1888,' p. 130.

## SUEZ CANAL.

**Levels of the Mediterranean and Red Seas.** When the French were occupying Egypt at the close of the eighteenth century, Napoleon I, then first consul, ordered a survey to be made across the Isthmus of Suez, with the idea of re-opening a waterway between the Mediterranean and the Red Seas. The rough and interrupted observations taken with this object, led to the belief that the level of the Red Sea was about 30 feet higher than that of the Mediterranean, which was disputed on theoretical grounds by the mathematicians Laplace and Fourier. It was not, however, till 1847 that careful levellings across the isthmus established the fact that the levels of the two seas are practically identical, except so far as they are affected by tides and wind.

The ordinary tidal variations in level amount, at Port Said, to 11 inches at springs, and 7 inches at neaps; and at Suez, to 4 feet 9½ inches at springs, and 2 feet 10 inches at neaps. The extreme observed differences, however, in the sea-levels at Port Said and Suez, resulting from the action of wind combined with the tidal rise or fall, amount to 4½ feet and 8⅔ feet respectively. The mean level of the Mediterranean is higher than that of the Red Sea in summer, reaching 1⅓ feet in September; and the reverse is the case in winter, amounting to a difference of 1 foot in January [1].

**Negotiations for the Suez Canal.** Probably it was a fortunate circumstance that the investigations directed by Napoleon did not result in the construction of a canal, for in the early part of the nineteenth century, the canal would have been made of comparatively small dimensions, and the Red Sea is not easily navigated by sailing vessels. Soon after the identity of the levels of the two seas had been established, Lesseps, in 1852, brought before the notice of

---

[1] 'Rapport de la Commission Consultative Internationale, 1884,' Paris, 1885, pp. 52–56, and plate 2.

Abbas Pasha, then Viceroy of Egypt, his proposal for cutting
a canal across the Isthmus of Suez, which he had long
pondered over during his diplomatic mission in Egypt, but
failed to gain the Viceroy's support. On the accession,
however, of Said Pasha in 1854, the scheme was at once
submitted again, and approved; and a survey of the district
afforded satisfactory indications of its feasibility. Political
opposition, nevertheless, and the expression of opinion by
some engineers that the project was impracticable, delayed
the starting of the enterprise, notwithstanding the highly
favourable report, in 1856, of an international Commission,
comprising three English engineers amongst its members, in
which the route for the canal was recommended which was
subsequently adopted [1]. At length, in 1858, a Company
was formed with a capital of £8,000,000, the estimated cost
of the undertaking; and the works were commenced in 1859.

**Route of the Suez Canal.** A depression of the land extends across the Isthmus of Suez, not far from the shortest
line between the Mediterranean and the Red Sea; and though
some circuitous routes were proposed, starting from Alexandria
and crossing the Nile at different points to reach Suez, the
far more direct route along the natural depression, entirely
to the east of the Nile delta, was naturally adopted for the
canal. The canal follows a fairly direct southerly course
from Port Said to Suez, except where it bends round to
avoid the higher ground at El Guisr, and flows into Lake
Timsah, and where it diverges towards the east to go through
the centre of the Bitter Lakes on its way to Suez (Plate 13,
Fig. 1). Starting from Port Said, the canal, along the first
43 miles, traverses lakes Menzaleh and Ballah, which are
marshy expanses of the Nile delta, lying for the most part
slightly below the sea-level, and separated from the Mediterranean by a narrow belt of sand. The canal then pierces the

[1] 'Isthmus of Suez Ship-Canal. Report and Plan of the International Scientific
Commission.' Paris, 1856, p. 39.

lower part of the ridge of El Guisr, which only rose 54 feet above sea-level along the line of the canal, and constitutes the deepest cutting, of about 80 feet, in the whole length (Plate 13, Fig. 2). After passing through Lake Timsah, the canal enters the cutting made through the somewhat lower Serapeum ridge separating Lake Timsah from the Bitter Lakes, the bed of which latter lakes, for about 10 miles, was nearly at the depth fixed for the bottom of the canal, and at about half that depth for most of the remainder of the distance. On leaving the Bitter Lakes, after a course through them of about 23 miles, the canal traverses the Chalouf ridge, and the lower land intervening between Chalouf and Suez. Lake Timsah and the Bitter Lakes were large dry hollows, from which the sea-water had at some previous period been evaporated, as indicated by the layer of salt covering their surface; and they were reconverted into lakes as the canal progressed, by water admitted from the sea through the cuttings.

**Description of the Suez Canal.** The canal, which is about 100 miles long between Port Said and Suez, was given a bottom width of 72 feet and a depth of $26\frac{1}{4}$ feet throughout, with side slopes of 2 to 1 in the deep cuttings of El Guisr and Serapeum, and $2\frac{1}{2}$ to 1 along the rest of the canal, up to a level of $6\frac{1}{2}$ feet below low water of spring tides between Port Said and the Bitter Lakes, and $4\frac{1}{2}$ feet below low tide from the Bitter Lakes to Suez. The widths of the canal at the top of the slopes, at these levels, were 170 feet through lakes Menzaleh and Ballah, 157 feet in the El Guisr and Serapeum cuttings (Plate 11, Fig. 32), and 182 feet from the Bitter Lakes to Suez; whilst above these levels, the widths were increased by berms or flatter slopes, so that, at the water-level, they were 328 feet, 190 feet, and 225 feet respectively.

Though the speed of transit was limited to an average not exceeding 6 miles an hour, the wash of the steamers tended to erode the banks which have been gradually protected by masonry; and dredging has been necessary to maintain the depth, which amounted, during the first twelve years, to an average of about 800,000 cubic yards annually. The canal was widened out at intervals of 5 or 6 miles, to enable vessels going in opposite directions to pass.

**Excavation for the Suez Canal.** The strata traversed by the excavations for the canal consisted chiefly of silt, sand, and clay of varying compactness, in which occasional thin layers of limestone and gypsum were encountered between Lake Ballah and the Bitter Lakes; but between the Bitter Lakes and Suez, though the material excavated consisted mainly of clay overlaid with sand, beds of marl, pebbles, and sandstone rock were met with in some places [1]. The total quantity of excavation amounted to about 98,000,000 cubic yards. At first the excavations were conducted by forced labour provided by the Viceroy, about 25,000 natives being supplied, who were changed every month. Soon after the accession, however, of Ismail Pasha early in 1863, the original concession was cancelled; and the forced labour was discontinued in 1864. This necessitated the importation of labourers from Europe; and as these were difficult to procure, and they were unused to the climate, mechanical appliances were introduced, as much as possible, to save manual labour.

The Egyptian labourers had excavated a cutting through El Guisr, down to about 6½ feet below sea-level, with a bottom width of about 40 feet, through which water from the Mediterranean could be admitted to Lake Timsah, so as to enable it to be deepened by dredging [2]. Small dredgers cut a narrow trench along each side of the canal through lakes Menzaleh and Ballah, depositing the material on the outer side to form a bank; and the canal was completed by larger dredgers removing the central portion and deepening the

---

[1] 'Rapport de la Commission Consultative Internationale, 1884,' plate 3.

[2] 'Travaux d' Exécution du Canal Maritime de l'Isthme de Suez,' A. Lavalley, Mémoires de la Société des Ingénieurs Civils, Paris, 1866, p. 485.

channel, the dredged material being deposited by long shoots on to the side banks (p. 93, and Plate 2, Fig. 7)[1]. Excavators, together with manual labour, were employed for widening the large cuttings. The cutting at Serapeum was partly effected by dredgers floating on lakes formed by filling wide depressions with water from the fresh-water canal; and the dredged material was deposited by hopper barges in the adjacent hollows, outside the line of the canal. As the water-level of these artificial lakes was several feet higher than the sea-level, the hollows were available for the deposit of the dredgings from hopper barges, which would otherwise have been dry; and the width of the cutting for the canal was considerably wider at this higher level, enlarging the area available for dredging and the waterway for the passage of the dredging plant. After completing the excavation to a depth of 26¼ feet, the dredgers were lowered to the sea-level, by letting out the surplus water into Lake Timsah, in order to complete the canal to its full depth; and the dredged material was deposited in Lake Timsah[2].

The excavation for the canal between deep water in the larger Bitter Lake and 2½ miles from Suez, was executed in the dry, both on account of the borings having revealed the existence of some rock under the shallow Bitter Lake, in the Chalouf cutting, and a little distance beyond, and also because the tidal fluctuations of the Red Sea would have impeded the dredging[3]. Dredgers, however, with long shoots and elevators, were employed for excavating the last 2½ miles of the canal, where the water was kept 6½ feet above the mean level of the Red Sea, by a bank across the entrance at Suez, and by raising the water-level with fresh water above the high-tide level.

**Dredging Plant on the Suez Canal.** For most of the

---

[1] 'Travaux d'Exécution du Canal Maritime de l' Isthme de Suez.' A. Lavalley, Mémoires de la Société des Ingénieurs Civils, Paris, 1866, plate 70.

[2] Ibid. pp. 514 and 515.     [3] Ibid. 1868, p. 606.

distance through the lakes, and along the low land near Suez, the dredged material could be deposited at the sides by long shoots; but for short distances, in a few places, the ground was too high for this method of deposit. Accordingly, an elevator was designed, consisting of two parallel iron girders, each about 158 feet long, placed at right angles to the canal, and rising with an inclination of 1 in 4 towards the shore, borne by framework resting at its centre on a movable support at the edge of the canal, and near its inner extremity on a pontoon floating on the canal[1]. The inshore half of the elevator formed a cantilever stretching over the land on which the bank had to be deposited, with its extremity rising to a height of 45 feet above the water-level. Boxes of 4 cubic yards capacity were placed upon a pontoon floating on the canal, and were filled by the dredger; and they were then lifted on to the elevator, and drawn up it by chains worked by a steam-engine; and on reaching the upper end, they tipped their contents on to the side bank, at a distance of 74 feet beyond the edge of the canal slope.

The dredging plant on the works at the end of 1866 comprised eighteen small dredgers, and fifty-eight large dredgers, twenty of which were provided with long shoots. The material raised by the dredgers without shoots was deposited by eighteen elevators with their attendant pontoons and boxes, and by thirty-seven large steam hopper barges which could go out to sea, and seventy-two smaller ones.

**Fresh-water Canal from Cairo to Suez.** As the route traversed by the Suez Canal was devoid of a supply of fresh water, and of means of communication for the conveyance of provisions, materials, and plant to the works situated in a barren, sandy desert entirely destitute of resources, the formation of a fresh-water canal to supply these necessities constituted an indispensable undertaking. This canal starts

---

[1] 'Travaux d'Exécution du Canal Maritime de l'Isthme de Suez, A. Lavalley, Mémoires de la Société des Ingénieurs Civils, Paris, 1866, p. 506, and plate 71.

from the Nile near Cairo, and on approaching Lake Timsah, a branch diverges, going past Ismailia to the ship-canal; whilst the main canal proceeds to Suez, in a direction approximately parallel to the ship-canal, with short branches to Serapeum and Chalouf (Plate 13, Fig. 1). The canal was made 26¼ feet wide at the bottom and 55¾ feet at the water-level, and 6½ feet deep. As the water in this canal is several feet higher than the sea-level, two locks were constructed near Ismailia to provide communication between the fresh-water canal and the ship-canal. There are three locks on the canal between Serapeum and Suez; and a lock at the end of the canal gives access to the Suez lagoons, through which the ship-canal passes. The canal is about 132 miles long between Cairo and Suez; and the branch to the ship-canal near Lake Timsah is about 5 miles long.

A supply of fresh water was provided along the ship-canal between Lake Timsah and Port Said, by pumping water from the fresh-water canal at Ismailia through two cast-iron conduits, 9 inches in diameter, laid along the ship-canal embankment as far as Port Said.

The dredgers and their attendant plant employed for the southern portion of the ship-canal, were brought from Port Said along the ship-canal to Lake Timsah, and were thence conveyed along the fresh-water canal to Serapeum and Suez.

The fresh-water canal was constructed by the Suez Canal Company; and in the original concession, the land that could be brought into cultivation by the water from the canal was to become the property of the Company. On the withdrawal, however, of the concession by Ismail Pasha, the canal and the 150,000 acres of land irrigated by its waters became the property of the State, though the Company retained the right of making use of the canal. The Company were indemnified for the loss of these rights, and for the withdrawal of forced labour, by the payment of £3,360,000, which money was employed for the prosecution of the canal works.

**Approaches to the Suez Canal at Port Said and Suez.** In order to preserve the dredged outlet channel of the canal from silting up, and to protect the approach to the canal at Port Said, a harbour, sheltering about 450 acres, was formed by two breakwaters, 9,800 feet and 6,233 feet long, starting from the shore on each side of the canal, 4,600 feet apart, and converging to an opening of 2,300 feet at the end of the shorter eastern breakwater (Plate 13, Figs. 3 and 4). The western breakwater was extended beyond the other, as the prevalent winds blow from the north-west, and the littoral current comes along the coast from the same quarter; but the eastern breakwater serves to protect the harbour in winter, when strong easterly winds occur. Owing to the difficulty of procuring stone, which had to be brought at first from quarries near Alexandria, the breakwaters were mainly constructed of mounds of concrete blocks formed of sand and Theil lime [1].

The drift of sand along the coast towards the east under the action of the waves, and the turbid littoral current bringing alluvium from the delta of the Nile, caused an advance of the foreshore on the outside of the western breakwater, and brought sand and silt into the harbour through the interstices in the concrete blocks. The progression, however, of the foreshore has now become very slow, and is not likely to endanger the approach to the harbour for a very long time; and as the spaces between the concrete blocks have become filled with sand, the introduction of sand into the harbour from the foreshore has ceased. Deposit, however, necessarily takes place to some extent in the harbour; and the outlet channel has to be maintained by dredging.

The outlet of the canal at Suez is very well protected in a land-locked bay of the Red Sea; and it is guided into deep water by an earthwork embankment on the north-west side, and by a mole, about 2,900 feet long, on the south-east side.

---

[1] 'Harbours and Docks,' L. F. Vernon-Harcourt, p. 218, and plate 5, fig. 7, and plate 6, figs. 10, 11, and 12.

**Opening and Cost of the Suez Canal.** The works of the canal were necessarily delayed for a time by the withdrawal of forced labour, the difficulty in procuring labourers from Europe, and the time required for constructing and bringing on to the site the large plant by which the excavations were eventually accomplished. The canal works were, however, sufficiently advanced in November 1869 for the formal opening of the canal, the filling of the Bitter Lakes, with about 1,962,000,000 cubic yards of water, having been commenced in the previous March from the Mediterranean, and in July from the Red Sea, and completed in October. A rocky layer discovered near the bottom of the Serapeum cutting, along a length of 280 feet, only a short time before the date fixed for the opening, limited the available depth at that time to 17 feet, and other portions of the canal had not been fully excavated; whilst the banks were still unprotected for long distances[1].

The cost of the canal, up to its opening, had reached £16,633,000, or rather more than double its original estimated cost[2]. Out of this sum, however, about £3,300,000 had been allotted to interest during construction and a sinking fund; whilst the amount received as indemnity for the loss of the privileges of the concession, and another payment by the Viceroy of £1,200,000 in 1869, helped to defray the cost of the works. On the other hand, the canal was given much smaller dimensions than had been originally proposed, to decrease its cost; whilst the expenditure, after the opening, in completing the canal, making improvements and enlargements, and protecting the banks, raised the total cost of the works to £19,845,000 by the end of 1878.

**Currents in the Suez Canal.** The influence of the small tidal rise in the Mediterranean at Port Said during spring

---

[1] 'Some Account of the Suez Canal,' J. F. Bateman, Proceedings of the Royal Society, 1870, vol. xviii. p. 132.

[2] 'Bulletin décadaire du Canal Maritime de Suez,' No. 22, December 1869.

tides extends, with diminishing effect, as far as the entrance to the Bitter Lakes, where there are no tidal currents. The larger tidal rise of springs at Suez diminishes more rapidly in passing along the canal, and almost dies out only a short distance beyond the entrance to the Bitter Lakes. The tidal oscillation at springs in the rest of the Bitter Lakes, where the tides from the two seas meet, amounts to only ¾ inch.

The northerly and north-westerly winds which prevail from May till October, raise the level of the Mediterranean at Port Said, and depress the Red Sea at Suez, producing a difference in level which attains about 1⅓ feet in September, creating an almost continuous current from Port Said to the Bitter Lakes, which reaches sometimes a maximum of 2 feet per second at the entrance, and 1½ feet at the far end of Lake Ballah[1]. During the same period of the year, the ebb current predominates between the Bitter Lakes and Suez, attaining sometimes a velocity of 4 feet per second when a strong north wind is blowing.

From November till May, the prevailing southerly winds raise the Red Sea above the level of the Mediterranean, producing a difference of level amounting to about 1 foot in January, and reversing the direction of the prevalent currents in the canal. Thus, the level of the Mediterranean being lower than Lake Timsah during this period, the outflow is almost continuous during January and February from Lake Timsah to Port Said, reaching a maximum velocity of 1½ feet per second near Lake Ballah, and nearly 2 feet at Port Said. For the same reason, the influx predominates from Suez to the Bitter Lakes from November to May, with an average velocity of 1½ feet per second, during which period a large volume of water flows from the Red Sea into the Bitter Lakes, which has been estimated at about 667,000,000 cubic yards. This large alternate influx and efflux through the canal has

---

[1] Comptes rendus de.l'Académie des Sciences, vol. lxxxvii. p. 142 ; and 'Rapport de la Commission Consultative Internationale, 1884,' p. 53.

reduced the deposits of salt in the Bitter Lakes, increasing thereby the depth of the lakes, and has entirely falsified the predictions that the canal would become a stagnant ditch.

**Traffic passing through the Suez Canal.** The number of vessels which passed through the canal in 1870, the first complete year after its opening, was 486, and rose to 1,082 in 1872; it has increased fairly steadily since then, with occasional falls for a year or two, up to a maximum of 4,207 in 1891 by an unusual rise, from which it dropped to 3,559 and 3,341 in 1892 and 1893, but began to rise again slightly in 1894, when it was 3,352 [1]. The net tonnage of all the vessels passing through the canal was 436,609 tons in 1870, 1,160,743 tons in 1872, 5,074,808 tons in 1882, a maximum of 8,698,777 tons in 1891, a fall to 7,659,068 tons in 1893, and a recovery to some extent in 1894, with 8,039,175 tons. The transit receipts were approximately £206,300 in 1870, £656,300 in 1872, £2,421,800 in 1882, a maximum of £3,336,600 in 1891, a drop to £2,826,600 in 1893, and a rise to £2,951,000 in 1894. The fall since 1891 seems rather to be attributable to the sudden leap up in 1891 not having been maintained, than to any marked cessation of the steady growth of traffic; for the returns for 1893, though lower than those for 1892 or 1894, were very materially higher than the returns for 1890. The mean net tonnage per vessel has increased very steadily from 898 tons in 1870 to 2,398 tons in 1894.

**Navigation of the Suez Canal by Night.** By 1883 the traffic on the Suez Canal had increased so much and so steadily, reaching in that year four times the tonnage of 1873, that experiments in lighting by electricity were commenced on the canal, with the object of increasing the accommodation and shortening the time of transit by conducting the navigation by night as well as by day. In 1885, electric lights were placed on high posts opposite all the passing places; and

---

[1] 'Suez Canal. Returns of Shipping and Tonnage, 1892, 1893, and 1894,' p. 8.

the entrances to the channels were marked by light-giving buoys, exhibiting a green light to starboard and a red light to port, between Port Said and Ismailia [1]. Aid was also provided for the pilots in steering vessels at night, by a projecting electric beam of light at the bows, and side and stern lights. By these means, a trial trip was made at the end of 1885; vessels of war and mail steamers were allowed to pass at night between Port Said and Ismailia in 1886; and in March 1887, the whole length of the canal was opened for navigation by night, in which year 393 vessels passed through by night, in an average period of $20\frac{1}{2}$ hours.

The average duration of the passage through the canal, which amounted to $48\frac{1}{2}$ hours in 1883, was reduced by the widening of the channel in the smaller Bitter Lake, by the flattening of the curve in Lake Timsah, the formation of Port Tewfik at the Suez end of the canal, and the introduction of the electric light, so that in 1887 the average time of transit was 34 hours [2]. In 1888, the number of vessels navigating the canal by night, as well as by day, increased to 1,608, with an increased time of transit of $22\frac{1}{2}$ hours, owing to the greater number; but the period of passage for all vessels was reduced to $30\frac{3}{4}$ hours. In 1889, the numbers passing by night increased to 2,457, reducing the average time of passage to $26\frac{3}{4}$ hours; and though since then, up to 1894, the number of vessels passing by night has fluctuated with the traffic, the percentage of these vessels has steadily increased, and the duration of the passage of vessels through the canal both by night and by day has decreased year by year. In 1894, the number of vessels passing through by night as well as by day was 3,180, or 94.8 per cent. of the whole number; and their passage, on the average, occupied only 19 hours 18 minutes, the average time of transit of all the vessels being 19 hours 55 minutes; whilst the average duration of the passage of

---

[1] 'Le Génie Civil,' vol. ix. p. 161.
[2] 'Suez Canal. Returns of Shipping and Tonnage, 1885, 1886, and 1887,' p. 3.

vessels navigating only by day, was reduced from 37 hours 57 minutes in 1888, to 31 hours 17 minutes in 1894.

**Widening and Deepening of the Suez Canal.** Even in 1876, owing to the considerable increase in the number and size of the vessels passing through the canal, the Company arranged to improve the waterway gradually in the worst places, and to enlarge the passing places and increase their number; and in 1882, it was determined to push on these works as rapidly as practicable. By 1884, however, the growth of the traffic led to the appointment of a Commission to consider the best way of doubling the waterway. Three schemes were investigated, namely, enlarging the existing canal, forming a second canal, and the enlargement of the canal from Port Said to the Bitter Lakes, with a second canal to Suez. The first scheme was naturally preferred, as affording greater scope for the traffic, facilitating the passage in an ample waterway, involving less erosion of the banks, and avoiding the blocking of the canal in the event of a vessel grounding (Plate 11, Fig. 32)[1]. The widths decided to be given to the canal, at a depth of 26¼ feet below low water of spring tides, between Port Said and the Bitter Lakes, were 213 feet in the straight portions, increasing to 246 feet in the middle of curves exceeding 1½ miles radius, and to 262 feet at sharper curves. Between the Bitter Lakes and Suez, the widths arranged were 246 feet in the straight portions, and 262 feet in the middle of the curves, which all exceed 1½ miles in radius, the greater widths in this portion of the canal being provided on account of the greater difficulties of navigating in the stronger tidal currents near the Suez end of the canal.

It was arranged that the widening through the deep cuttings of El Guisr and Serapeum should be all effected on the African

[1] Minutes of Proceedings Institution C. E., vol. lxxvi. p. 238; and 'Rapport de la Commission Consultative Internationale, 1884,' pp. 75 and 110, and plates 4 and 5.

side of the canal; that through Lakes Menzaleh and Ballah, the widening should be mainly on the side towards Asia ; and that the widening between the Bitter Lakes and Suez should be carried out equally on both sides. The slopes are to be protected near the water-line by stone pitching set in cement through Lakes Menzaleh and Ballah, and from the Bitter Lakes to Suez. Plantations of reeds and tamarisk will be used, as at present, to protect the slopes near the water-line, through the enlarged cuttings in firm soil of El Guisr and Serapeum.

The Commission decided that the canal should be deepened in the first instance to 28 feet below low water of spring tides; and that eventually a minimum depth of 29½ feet should be provided throughout. The total excavation required for widening and deepening the canal to the full extent, was estimated at about 90,420,000 cubic yards; and the estimated cost is £8,118,000.

The work of enlargement is being carried out ; and the rock in the Chalouf cutting, and towards Suez, is being shattered under water by rock-breaking rams (p. 91, and Plate 2, Fig. 10). As the widening proceeds, the navigation of the canal will be more and more facilitated, and the time of the passage reduced; and it is proposed, when the work has been finished, to allow vessels to go through the canal at a speed of about 9 miles an hour. Already vessels of greater draught than formerly are enabled to navigate the canal ; for whereas up to 1889 the maximum draught was limited to 24 feet 7 inches, in 1890 vessels drawing 25 feet 7 inches were allowed to pass through ; and in 1894 the number of vessels drawing over 24 feet 7 inches was 172, of which 66 had a draught of 25 feet 7 inches.

**Remarks on the Suez Canal.** The chief difficulties encountered in the construction of the Suez Canal were the large amount of earthwork that had to be removed by excavation or dredging, the absence of all resources along the site of the works, so that even the supply of fresh water for the workmen involved the execution of a long subsidiary canal, the necessity of providing a large number of dredgers and excavators, with their attendant plant, to cope with the excavations on the withdrawal of forced labour, the deficiency of the means of transport to the various sections of the canal works, and the large amount of capital that had to be raised to complete the works. Moreover, a canal intended solely for through traffic, like the Suez Canal, is not available for trade until it has been opened from end to end. In other respects, however, the conditions were peculiarly favourable, for the route follows a natural depression across the isthmus, along which the surface of the ground was close to the water-level of the canal for nearly half the distance, and several feet below it through the Bitter Lakes (Plate 13, Fig. 2); no land had to be acquired or rights purchased; no locks or bridges had to be erected across the canal; and the only constructive works of importance were the breakwaters and basins at Port Said. The great present financial success of the canal has obliterated its early financial difficulties; but it is well to bear in mind, in view of similar undertakings with less grand possibilities before them, that whilst the original estimate was intended to provide a canal wide enough for two vessels to pass at any point, the canal, as executed, was only made wide enough for the passage of a single vessel, except at the passing places, at an expenditure amounting to double the estimated cost.

This remarkable undertaking, which has transformed the conditions of ocean-going traffic between Europe and the East, will ever be associated with the name of Lesseps, the originator and leader of the enterprise, whose diplomatic training and official position in Egypt aided him greatly in starting his scheme. The success, however, of the canal as a bold engineering undertaking, was due to Voisin Bey, the engineer-in-chief of the works, Mr. Lavalley, the contractor who organized the dredging and other mechanical appliances for expediting the work, and their engineering assistants.

A notable peculiarity of the Suez Canal is that it is held by the Company under a concession lasting for ninety-nine years; and when at the end of this term of years, which will expire in 1968, the canal reverts to the State, Egypt will come into possession of a magnificent inheritance, which it may be hoped will be utilized for the advancement of the prosperity of that country.

The remarkable success of the Suez Canal, and the large traffic which was soon attracted to it, naturally led to proposals for cutting ship-canals across other barriers to navigation; but unfortunately the differences in the physical conditions, or in the prospects of traffic, were not adequately considered; and none of the schemes for isthmian canals, which were subsequently started, have resulted in a similar success.

## CORINTH CANAL.

The cutting of a waterway across the Isthmus of Corinth appears to have been first suggested by Periander, about 600 B. C.; and the idea was revived subsequently from time to time, till at last, in the reign of Nero, trial pits were sunk in places across the narrow part of the isthmus, and the works actually commenced, which were, however, abandoned on the death of their promoter Nero. Traces of the excavations for the canal were found when surveys for the new works were made along the same line; and the trial pits were resorted to for ascertaining the nature of the strata when the canal works were commenced in 1882.

**Object of the Corinth Canal.** When Athens was the centre of the civilized world, the connection of the Gulf of Corinth with the Gulf of Ægina, by a waterway across the Isthmus of Corinth, would have possessed a very great importance; and even in the time of the Roman emperors, the shortening of the journey between Rome and Athens would have been very advantageous. Though, however, the main routes of

commerce have, since that period, been to a great extent
diverted into other directions, and the introduction of steam
and the increase in the size of vessels have greatly reduced
the dangers and delays of a stormy voyage round Cape
Matapan, the route through the Isthmus of Corinth shortens
the distance between the Adriatic and the Black Sea ports
by 212 miles, and between the western Mediterranean and the
Black Sea ports by 110 miles, in respect of the vessels
passing through the Straits of Messina.

**Description of the Corinth Canal.** In 1881 the Hungarian
General Türr, having relinquished his concession for forming
the Panama Canal to Lesseps, obtained a concession for cutting
a canal through the Isthmus of Corinth; and the works were
commenced in 1882.

The width of the isthmus at the site of the canal is slightly
less than four miles, the actual length of the canal being only
3 miles 7½ furlongs; and deep water is reached about 200 to
300 yards from the shore at each end [1]. Though, however, the
isthmus is very narrow, the ground rises so rapidly towards
the centre, that the depth of the cutting for the canal averages
190 feet for a length of 2⅗ miles; and a maximum height
of 287 feet above the level of the bottom of the canal is
attained (Plate 13, Figs. 7 and 8). The canal is a straight
cutting throughout; and the dimensions determined upon
were a bottom width of 72 feet, and a depth of 26¼ feet below
the lowest sea-level, like the Suez Canal, and side slopes
of from 2 to 1 to 1 to 1 in soft soil, and 1 in 10 through
rock (Plate 11, Fig. 29). The depth was eventually proposed
to be carried to 27 feet 10 inches; but the available depth
of the canal, as executed, appears to be three or four feet
less than this.

Harbours were formed at both ends to protect the entrances
to the canal, by constructing breakwaters of mounds of rubble
stones, and dredging the sheltered channel to the requisite

[1] 'Annales des Ponts et Chaussées,' 1888, (2) p. 458, and plates 23 and 24.

depth. Two converging breakwaters, 1,310 feet and 1,640 feet long respectively, form the harbour in the Gulf of Corinth, starting from the shore 1,150 feet apart, and leaving an entrance 262 feet wide between their extremities ; whilst the approach to the canal in the Gulf of Ægina has been adequately sheltered by a single breakwater on the northern side.

Only one bridge, of 262 feet span, crosses the canal, at about the centre of the isthmus, at a height of 141 feet above the water-level, which, carrying both a railway and a roadway, maintains the communication between the Morea and the mainland. The Canal Company has also to maintain a free ferry service across each end of the canal.

**Construction of the Corinth Canal.** During the first three years, the approaches to the canal in both gulfs were dredged, the breakwaters were constructed, the formation of the canal was commenced at each end by dredging in the alluvial soil, and the upper portion of the rock cutting was begun by excavating on the surface, and by sinking shafts in the line of the canal, which were connected at the bottom by a gallery in which the wagons were loaded with the excavated materials through side drifts, and were then conveyed away by loco-motives to the place of deposit. The lower rock excavation was intended to be removed by special dredgers, after the rock had been shattered by boring deep vertical holes and blasting with dynamite. This system, however, of blasting was not successful ; and in 1885, the method of excavation adopted for the upper part of the cutting was extended to the lower portions ; and locomotives conveyed trains of wagons from a series of galleries, at different heights, to their respective spoil banks.

As the excavations progressed, it became manifest that the soundness of the sides of the pits sunk in Nero's time to only a portion of the depth to which the cutting had to be carried, upon which reliance had been placed in designing the side slopes in rock at 1 in 10, did not afford a true indication of the nature of the strata. Numerous faults were disclosed in opening out the cutting, producing dislocations and irregularities in the strata, which, however, in spite of these disturbances, have fortunately remained nearly horizontal. Blue marls, consisting of beds of clayey limestone, were found underneath the upper strata of conglomerates and limestone; and these marls, though fairly uniform in composition, and becoming harder when exposed to the dry climate of Greece, were disintegrated by contact with water[1]. These disturbances in the strata, resulting in some slips as the excavations were carried down, made it manifest that a slope of 1 in 10 could not be maintained in most parts of the cutting. Accordingly, it was decided, in 1887, that, except along a length of ¾ mile where the rock was compact, the rocky slopes above the water-level should be flattened to at least 1 in 5, and even to about 2 in 3 where the strata were less firm, and that the slopes from the bottom of the canal to a little above the water-level should be reduced to 1 in 4 and lined with masonry along the central 2⅗ miles, involving a considerable increase in the necessary expenditure.

The excavation for the canal when the works were commenced was estimated at about 12,430,000 cubic yards; and between 1882 and 1889 a total of 10,752,000 cubic yards were removed, the greatest amount excavated in a single year reaching 2,355,000 cubic yards in 1888. The flattening of the slopes, and the excavation for the lining of the sides of the canal entailed an additional excavation of about 2,350,000 cubic yards, raising the total excavation to 14,780,000 cubic yards. Accordingly, when the works were brought for a time to a standstill from lack of funds in 1889, there remained about 4,000,000 cubic yards of excavation to complete the cutting for the canal, as well as the principal portion of the masonry lining and some minor supplementary works.

---

[1] 'Le Génie Civil,' 1890–91, vol. xviii. p. 131.

**Completion and Cost of the Corinth Canal.** The original concession stipulated that the canal should be completed in 1888 ; and the capital of the Company for the construction of the canal had been fixed at £1,200,000. In 1886, the more accurate knowledge of the strata obtained by the deepening of the cutting, demonstrated that the work could not be accomplished within the allotted time, or for the estimated cost. Accordingly, in 1887, an extension of time to the end of 1891 was obtained from the Greek Government ; and the capital was doubled. As, however, only one-third of the new capital was subscribed for, an arrangement was made with the Société du Comptoir d'Escompte of Paris to provide the funds necessary for the completion of the works, in consideration of the allotment to it of the remaining two-thirds of the new shares created, and the selection of a contractor being entrusted to the Society[1]. The work was steadily prosecuted by the new contractor up to March 1889, when the failure of the Society, and the consequent cessation of payments, necessitated the stoppage of the works, and led to the liquidation of the Canal Company. A new company, however, was constituted, under the title of Société Hellénique du Canal de Corinthe, which arranged a fresh contract for the completion of the works, and obtained a further prolongation of the period within which the canal was to be opened.

The works were eventually completed in 1893, the inauguration taking place in August 1893 ; and the canal was actually opened for traffic in the following month. The total cost appears to have amounted to about £2,750,000 ; but, owing to the changes in the constitution of the Canal Company and the conduct of the works, it is difficult to arrive at the precise cost of the undertaking.

**Navigation and Traffic on the Corinth Canal.** The sea-level

---

[1] 'Le Canal de Corinthe,' A. Saint-Yves, Manchester Inland Navigation Congress, 1890, p. 6.

in the Gulf of Corinth, under the influence of north-westerly winds,varies during the highest springs between 4¾ inches above and 15¾ inches below the mean; and in the Gulf of Ægina, which is more sheltered from the wind, and where the tidal oscillations are very slight, the variations above and below the mean are only from 1½ to 4 inches. The difference in the level of the sea at the two ends of the canal only on exceptional occasions reaches 1⅔ feet; but the current in the canal is generally a little over a mile an hour, and occasionally reaches about 3½ miles an hour, when great care is needed to prevent vessels touching the steep side slopes of the canal lined with masonry, and thus injuring their plates in navigating the canal. The navigation of the canal by large vessels is also impeded by the small cross-section of the canal with its steep sides, in comparison with the Suez Canal and other ship-canals with flat slopes and berms (Plate 11, Figs. 20, 25, 26, 29, and 32). The speed of vessels in passing through the canal is limited to about 7 miles an hour. The traffic through the canal hitherto has been disappointing ; for though the saving in time has been estimated at 17 hours for steamers availing themselves of the canal, and the Lusitania, 415 feet long and drawing 22 feet of water, passed safely through it in 1894, the principal lines of steamers trading in these regions have not adopted this route.

**Remarks on the Corinth Canal.** The construction of the Corinth Canal, though due to the impulse given to such schemes by the success attending the Suez Canal, exhibits a striking contrast to the Suez Canal, both in the nature of the undertaking and its importance. The Corinth Canal, though very short, involved a cutting three and a half times the depth of the deepest cutting in the Suez Canal ; and though the rock constituting the main portion of the ridge to be traversed, enabled steep slopes to be adopted, it was too dislocated and variable in nature to allow of the exceptionally steep slopes originally contemplated. The steep sides, moreover, combined with a narrow bottom width and a current, render the navigation of the canal somewhat awkward. The possible traffic through the Corinth Canal is necessarily limited in extent, though the actual tonnage of the vessels passing round Cape Matapan was estimated at 12,000,000 tons annually, of which it was assumed that 5,900,000 tons would pass at the outset through the canal. This has proved a very excessive computation, which a consideration of the traffic on the Suez Canal during the first few years should have rendered evident. The Suez Canal, moreover, is one of the great highways of trade, and has greatly shortened the voyage between Europe and the East; whereas the Corinth Canal only affords a somewhat shorter passage between certain ports on the Mediterranean and the Black Sea, together with a portion of the west coast of Asia Minor. The advantage, however, of this shortening of the route is seriously impaired by the canal not being always accessible to vessels of 23 feet draught, and especially by the narrowness of the canal, with its hard steep sides, and the current through the canal, which occasion difficulties in navigating it. The traffic on the Corinth Canal will probably increase gradually, if it is found by further experience that the canal can be traversed safely with ordinary care; but impediments to navigation which may appear small, suffice to deter the owners and captains of vessels from using a route which affords only a moderate saving in time. The bottom width, indeed, of 72 feet, with almost vertical sides, is too narrow for a ship-canal intended for the passage of large ocean-going vessels; but unfortunately no funds are available for enlarging the Corinth Canal.

**Perekop Canal.** Another isthmian canal, of comparatively minor importance, has been recently carried out in Europe, traversing the Isthmus of Perekop which unites the Crimea to the mainland of Russia. This canal connects the Sea of Azov with the northern part of the Black Sea, thereby shortening the distance between the ports of Odessa and Kherson and the ports of the Sea of Azov by about 125 miles. Though the

Isthmus of Perekop is only 5½ miles wide at its narrowest point, the canal has had to traverse a long extent of shoal water before reaching the Sea of Azov, so that the total length of the artificial waterway is about 73½ miles. The canal has been given a depth of only 14 feet, as the Sea of Azov is shallow. Coasting vessels can now pass with safety between the Don and Dnieper, avoiding the stormy passage round the south of the Crimea; and Odessa has been placed in direct communication with the coaling port of Marioupol on the Sea of Azov. The canal will also form a link in the waterway between Odessa and the Caspian, whenever the Don and Volga are united by the proposed canal.

# CHAPTER XXV.

## INTEROCEANIC SHIP-CANALS (*continued*).

Importance of severing the Barrier formed by the Isthmus of Panama. Routes proposed for Canals across the Isthmus of Panama—Nicaragua, Panama, San Blas, Atrato, Tehuantepec Ship-Railway. *Panama Canal* : Preferred to other schemes; Route, and estimated Cost; Description, Dimensions of Canal, cuttings, approaches, provisions for Floods of river Chagres ; Excavation, increased amount, nature of Strata; Progress of Works, Contract, increased Cost, state in 1887 ; Introduction of Locks to reduce Excavation; Stoppage of Works, estimated Cost of Completion, Schemes proposed ; Remarks. *Nicaragua Canal* : Preferred by Americans ; Line selected, general course, length ; Dimensions of Canal and Locks, lifts of Locks ; Excavation, nature of Strata; Estimated Period of Construction, and Cost ; Commencement of Works ; Stoppage ; Remarks. *Comparison of Panama and Nicaragua Canals* : Relative Advantages and Disadvantages ; Problems involved ; Future Prospects, possible Intervention of United States Government. *Various schemes for Ship-Canals* : Minor Importance ; Isthmian Ship-Canal Schemes ; Ship-Canals proposed for Inland Cities ; Concluding Remarks.

THOUGH several schemes for shortening the routes of ocean-going traffic, by the construction of ship-canals across comparatively narrow stretches of land separating two seas in different parts of the world, have been brought forward from time to time, the Isthmus of Panama is the only remaining barrier, the severance of which possesses an interest at all comparable to the piercing of the Isthmus of Suez. The westward traffic from Europe to the eastern portion of Asia, and Australasia, has to make the very circuitous and stormy voyage round Cape Horn; and a waterway through the Isthmus of Panama would not only shorten very much the route of this traffic, but would also provide fairly direct access by sea between the ports of the United States on the Atlantic and Pacific coasts.

**Routes proposed for Canals across the Isthmus of Panama.** The Nicaragua route is the most northern of the routes proposed for a canal; and crossing the isthmus at a comparatively wide part, it would have a length of about 170 miles.. By utilizing, however, Lake Nicaragua as the summit-level of the canal, and damming up the river San Juan on the Atlantic slope, a great portion of the route would be open navigation, which vessels could traverse at a good speed; and the cost of the works might in this manner be brought within reasonable limits (Plate 13, Figs. 9, 10, and 11). Locks near each extremity of the canal would enable vessels to surmount the difference of about 110 feet between mean sea-level and the highest level of the lake; and artificial harbours on the two coasts would protect the entrances to the canal

The Panama route traverses a narrow part of the isthmus, following approximately the course of the Panama Railway which provides an overland route; and it was proposed to cut a level canal from ocean to ocean, 46 miles long, with a harbour at each end (Plate 13, Figs. 5, 6, and 11).

The San Blas route, starting from the Atlantic Ocean near Cape St. Blas, and entering the Bay of Panama to the east of Panama, would enable a level canal to be formed with a length of only about 40 miles, along the last ten of which the river Bayano would be utilized; but it possesses the very serious disadvantages of necessitating a tunnel 7 miles in length, and a bad site for a harbour on the Pacific coast.

The most southern scheme is the Atrato route, which was designed to follow the river Atrato, from its outlet in the Gulf of Uraba on the Atlantic, for 149 miles, and to reach the Pacific by a canal, 31 miles long, following the valley of the river Napipi. This canal, however, would have required some locks, and a tunnel 5½ miles long to pierce a high ridge.

The Tehuantepec Ship-Railway was another scheme proposed some years ago for enabling vessels to traverse the isthmus at a narrower part to the north-west of the Nicaragua. route (Plate 13, Fig. 11).

Only two of these schemes have taken practical shape, namely the Panama Canal and the Nicaragua Canal, for the construction of both of which companies were formed, and works commenced, which were subsequently stopped for want of funds.

## PANAMA CANAL.

The completion and commercial success of the Suez Canal naturally directed attention to the cutting of a canal across the barrier which separates the Atlantic and Pacific Oceans. At an International Congress held in Paris in 1879, the relative merits of the four canal schemes referred to above, as well as one or two others, were considered; and after long deliberation, the Panama scheme was adopted by a large majority, and Lesseps was requested to assume the lead in the enterprise [1]. Preference was given to the Panama route by Lesseps and others, on account of it being the only route along which a canal without locks or a tunnel appeared practicable; and this route was also considered superior to the Nicaragua route, because this latter route is much longer, and traverses a district more subject to earthquakes than Panama.

**Route and estimated Cost of the Panama Canal.** The route of the Panama Canal had been already surveyed; but fresh surveys were made in 1879–80, and borings taken. The line selected for the canal starts from Colon, in the Bay of Limon, on the Atlantic, and follows the valley of the river Chagres from Gatun to Obispo, where it enters the valley of the river Obispo, along which it runs to near Emperador, where it leaves this valley to cross the lowest part of the Culebra ridge, from whence it passes into the valley of the river Grande and enters the Pacific Ocean near Panama (Plate 13, Figs. 5 and 6). A concession was granted by the Columbian Government in

---

[1] 'Congrès international d'études du Canal interocéanique, 1879.'

1880 for the construction of the canal, which was to be completed in eight years; and the works were commenced in 1881. An engineering Commission estimated the cost of the works at £33,700,000; but as it was considered by the promoters that an excessive price had been put down for the earthwork, the exact nature of which had not been ascertained, the capital of the Company was fixed at £24,000,000. The Panama Railway was purchased by the Canal Company in 1882 for £3,730,000, as its proximity to the line of the canal rendered it very convenient for the conveyance of plant, materials, and stores to the various sections of the canal works.

**Description of the Panama Canal.** The canal, 46⅜ miles in length, which was proposed to be carried across the isthmus at a uniform level, to avoid the delays involved in the passage of vessels through locks, was designed to be given a depth of 27⅝ feet below the sea-level of the Atlantic through soft soil, and 29½ feet through rock. The widths determined upon for the canal, in soft soil, were 72 feet at the bottom and 160 feet at the water-level, with slopes of 1½ to 1, and in rock, 78¾ feet and 91⅛ feet respectively, with slopes of 1 in 4½, following approximately the original section of the Suez Canal, with passing-places provided at intervals (Plate 11, Figs. 27 and 28).

The line of the canal follows a south-easterly direction from the Bay of Limon to the Bay of Panama; and several curves had to be introduced in its somewhat winding course to enable the canal to follow the valleys, and thus keep down the amount of excavation; whilst the Panama Railway is crossed twice by the canal, namely near San Pablo and at the Culebra ridge (Plate 13, Fig. 5). The depth of cutting required for a canal at the sea-level, reaches 346 feet at the summit of the Culebra ridge; and the necessary cutting averages about 120 feet in depth for a length of ten miles across the central mass (Plate 13, Fig. 6).

As the canal opens into fairly sheltered bays at the two ends, the formation and protection of its approaches presented no special difficulties; and the only works required for this purpose were the dredging of a channel for some distance across the foreshore at each extremity to reach deep water, and the construction of a breakwater, about $1\frac{1}{4}$ miles long, on the Atlantic side to protect the Bay of Limon from northerly winds. The practical absence of tide on the Atlantic side, with a rise at springs of only about $1\frac{1}{2}$ feet, enabled the canal to be in free communication with the sea at Colon; but as the rise at spring tides on the Pacific coast reaches $21\frac{1}{3}$ feet, it was necessary to provide for the construction of a regulating lock at that end, to prevent the creation of strong tidal currents in the canal when the tidal range is large. The route of the canal frequently crosses the winding river Chagres in following its valley on the Atlantic slope, necessitating the construction of numerous diversions; and as the floods of this torrential river rise very rapidly and to a great height in that rainy district, it was essential to prevent the floods of the river rushing into the canal and causing damage, which it was proposed to accomplish by forming a large dam, about 130 feet high, across its valley near Gamboa, to impound the flood waters which were to be discharged into the sea through artificial channels away from the canal. The control of the floods of the Chagres, and the immense volume of earthwork to be removed from the exceptionally deep cuttings, constituted the main difficulties to be encountered in the construction of the Panama Canal.

**Excavation for the Panama Canal.** As the excavation is the principal item in the cost of isthmian ship-canals, it is of the utmost importance that the nature of the strata to be traversed should be ascertained as closely as possible before estimates are made or the undertaking commenced. In the case, however, of the Panama Canal, as for the Corinth Canal, reliance was placed at the outset on insufficient data;

whereas the exceptionally great amount of excavation required for cutting through the Isthmus of Panama, rendered the closest investigations of paramount importance for this particular work.

At first it was assumed that about 60,000,000 cubic yards of excavation would suffice for the construction of the canal, as it was supposed that the rock approached the surface, and that therefore a steep slope could be to a great extent adopted for the deep cutting through the Culebra ridge. Subsequently, fresh investigations showed that a considerable thickness of clay overlaid the rock; and before the works were commenced, the estimated amount of excavation had been raised to about 100,000,000 cubic yards, about half of which it was considered would be rock. Later on it was realized that an excavation of at least 176,500,000 cubic yards would be necessary for the formation of the canal. The strata consist of alluvial formations at each end, and of a thick layer of clay, interspersed with boulders, overlying disintegrated schists and tufa, with solid conglomerates, limestone, and sandstone at different parts below. The solid rock was found to extend at a very variable level across the central ridge to be traversed by the canal, and also to crop up in the lower part of the cutting for about 5 miles near Bohio-Soldado, as indicated by the dotted line on the longitudinal section (Plate 13, Fig. 6).

**Progress of the Panama Canal.** At the beginning of 1881, a contract was entered into with Messrs. Couvreux and Hersent for the construction of the canal for a total sum of £20,480,000, exclusive of the cost of the regulating lock at Panama and the general expenses of administration of the Company; and the works were commenced by these contractors[1]. In two years the contractors terminated their contract, in accordance with the option given them under a provision in the contract which had not been publicly

---

[1] 'Bulletin bimensuel du Canal interocéanique,' p. 315.

stated ; and the Company had to proceed with the work independently. The canal was opened out at each end by dredgers ; and numerous excavators attacked the cuttings at various parts of the line. It, however, soon became evident that the disturbance of the soil in that tropical region increased the unhealthiness of the climate, that the cost of the excavation had been considerably under-estimated, and that the upper stratum of clay, when exposed to the tropical rains by the opening out of the cuttings, became disintegrated and slipped, so that it could not stand at anything like the slopes contemplated. The original capital of the Company, accordingly, became exhausted before long, and additional funds were raised on increasingly onerous terms ; and it has since been disclosed that a considerable proportion of the capital obtained was misapplied, and not expended on the works.

At length, in 1887, the promoters became convinced that a canal at the sea-level would occupy a much longer period in construction than had been imagined, and would cost far more than the funds they could hope to raise. At this period, the canal had been nearly completed along the first 10 miles from Colon ; and more than half the total excavation along this section, 16⅓ miles long, had been accomplished. At the Panama end also, the last 4½ miles of the canal were almost completed ; whilst about one-third of the excavation of this section, 10⅔ miles long, had been carried out[1]. In the three central sections, however, 19 miles in length altogether, in which the bulk of the excavation was situated, the progress had been much less proportionately, especially in the deep cutting of the Culebra section, where, owing to large slips in the clay and disintegrated schist, in which the cutting could with difficulty be formed with very flat slopes, not quite a thirteenth of the contemplated excavation for a level canal had been effected. Altogether, only 39,200,000 cubic yards had up to that

[1] 'Le Génie Civil,' 1888, vol. xii. p. 178.

time been excavated, out of the estimated total of 176,500,000 cubic yards, or between a fourth and a fifth of the whole, though the Company had already come nearly to the end of its resources, in spite of the additions to its capital. The work, however, was being prosecuted with activity by the aid of seventy-two excavators, twenty-four dredgers, a number of locomotives and wagons, some transporters, and various other plant, an average of about 1,570,000 cubic yards being excavated per month.

**Introduction of Locks on the Panama Canal.** In order to reduce the amount of excavation, and to expedite the opening of the canal, it was decided in 1887 to introduce locks provisionally on each slope, for the working of the canal, till funds could be procured for completing the original scheme. Five locks were to be introduced on the Atlantic slope, as shown on the longitudinal section, two with a lift of 26¼ feet and three with a lift of 36 feet, raising the summit-level through the Culebra ridge to 160½ feet above the level of the Atlantic (Plate 13, Fig. 6). Four locks also, placed much closer together on the steeper Pacific slope, with one lift of 26¼ feet and four lifts of about 36 feet, were to surmount the difference of 170½ feet in level between low water of spring tides in the Pacific and the summit-level. All the locks were designed to afford an available length of 590 feet, and a width of 59 feet; and they were to be closed at each end by rolling, counterpoised caissons[1]. Arrangements were made in the designs for the rapid filling and emptying of the locks, which were to be supplied with water from the rivers Chagres, Obispo, and Grande; and it was estimated that vessels would be.able to traverse the canal in 17½ hours, allowing an hour for the passage through each lock. This modification in the levels of the canal reduced the requisite excavation for completing the canal from 137,300,000 cubic yards to 52,300,000 cubic yards, which, at the rate of progress in 1887, could have

[1] 'Le Génie Civil,' 1888, vol. xii. pp. 181 and 244.

been accomplished in three years. Notwithstanding, however, this great reduction in the earthwork, the estimated cost of the canal had risen to £65,500,000.

Stoppage of the Panama Canal Works. The works were prosecuted on this basis in 1888, as these radical changes in the scheme raised hopes for a time that the canal might thereby be opened for traffic within a reasonable period, and at a cost that might be defrayed. As the available funds, however, became exhausted in the course of the year 1888, the works had to be stopped; and the Company went into liquidation at the beginning of 1889. At that period, about 52,500,000 cubic yards had been excavated, leaving a balance of about 39,000,000 cubic yards of earthwork remaining to be done. The expenses connected with the undertaking had reached a total of £52,078,500, of which only £35,160,000 had been expended at Panama, including the purchase of the Panama Railway, and £98,200 paid to the Colombian Government, nearly £17,000,000 having gone in the costs of administration in Paris, interest on capital during construction, and extensive misappropriations of funds.

A Commission of inquiry, appointed in 1890, reported that the completion of the canal would necessitate a further expenditure of £36,000,000. It was proposed to dispense with the dam for impounding the Chagres floods, which presented considerable difficulties, and to place one of the reaches on the Atlantic slope at such a level as to form a large lake, into which the river could be admitted without impeding the navigation of the canal, except possibly for a short period during exceptional floods. Various other schemes and estimates have been brought forward for the completion of the canal : in one scheme it is proposed to substitute inclined planes for locks, as more economical [1]; and in another, a ship-railway has been suggested for surmounting the Culebra

[1] 'Étude pour l'Achèvement du Canal de Panama,' J. de Coene, Rouen, 1893.

ridge, whereby the large central cutting would be almost entirely dispensed with [1]. Efforts have been made from time to time to form a new Company, and raise fresh capital for completing the canal; and in the autumn of 1894, a new Company was formed, with a capital of only £2,600,000, in order to keep in force the concession which had been prolonged, and to institute more thorough investigations into the feasibility and cost of completing the canal with locks

**Remarks on the Panama Canal.** The successful completion of the Suez Canal, demonstrating conclusively the fallacy of the predictions of failure, and of the reports as to the impracticable nature of the work, led persons to minimize the difficulties in the construction of the Panama Canal, a work apparently somewhat similar in character, to be executed under conditions which, though different, seemed in some respects more favourable, and possessing similar prospects of becoming one of the most frequented routes for ocean-going trade. The site of the Panama Canal, with tropical vegetation and an abundance of water, seemed preferable at first sight to the sandy desert traversed by the Suez Canal; fairly accessible harbours existed at the two ends of the Panama Canal; the width of the isthmus at Panama is less than half that of Suez; and the Panama Railway offered convenient access to the works, which had to be provided at Suez by the construction of the fresh-water canal. Experience also had been gained, in the construction of the Suez Canal, of the methods by which large excavations could be rapidly effected, and the earthwork transported to the sides by powerful mechanical appliances. All these seemingly favourable conditions doubtless greatly influenced the framers of the original estimates for the Panama Canal; though some of the engineers who attended the Congress of 1879 urged the need of caution, and expressed the opinion that a far more searching investigation of the nature of the

[1] 'Achèvement économique du Canal de Panama,' Paris, 1892.

strata to be traversed, and a more careful consideration of the
other conditions were essential before a decision could be
arrived at as to the practicability and cost of constructing
a level canal across the isthmus. The public also were most
naturally misled by the announcement that a well-known
firm of contractors had undertaken to construct the canal
for a definite contract sum, which eventually proved illusory.

In reality, the Panama Canal works were subject to many
serious disadvantages, from which the works at Suez were
exempt. The funds for the Panama Canal had to be wholly
provided by the public; whereas at Suez, half the original
capital for the canal was subscribed by the Viceroy of Egypt,
who also granted various other valuable concessions.

The climate of Panama proved very unhealthy and ener-
vating; and the rate of wages there is very high. The excavations at Panama, though comprised in a much shorter length
than at Suez, are far less accessible for dredging; the much
greater depth of the cuttings renders the removal of the earthwork much more difficult and costly; and the upper strata
at Panama consist of treacherous clays and schists. Moreover, instead of the dry, rainless climate of Egypt, very
continuous tropical rains at Panama disintegrated the slopes
of the cuttings, and caused frequent slips, so that the excavations proved much more costly at Panama than in Egypt,
as careful investigations of the strata at the commencement
would no doubt have indicated. In Egypt, there were no
rivers or streams to interfere with the works; whereas the
river Chagres, with its excessive floods, offers a most serious
obstacle to the construction of the Panama Canal. Indeed,
except as regards access to the works, the supply of water,
and previous experience, the conditions at Panama are far
more unfavourable than they were at Suez; but the favourable
conditions were brought prominently forward, and the others
were disregarded.

The Panama Canal was commenced without any adequate

knowledge of the nature of the strata, or at all a sufficient
consideration of the other conditions of the work; and the
capital, by a reduction of the original estimate formed on
inadequate data, was fixed at an amount only slightly exceeding the cost of the Suez Canal, though the excavation required
for a level canal has proved to amount to about double
the excavation for the Suez Canal, with a much larger
proportion of rock. The result of thus hastily embarking
upon an undertaking of such magnitude has been that, after
nearly eight years' work, and after exhausting all the available
resources of the Company, the estimated expenditure for
completing the canal with locks, and thereby reducing the
excavation by about one half, amounts to rather more
than the original estimate for a level canal, which was regarded
as excessive, and half as much again as the original capital
of the Company. No stronger warning could be afforded
of the folly of starting a colossal engineering work without
a very exhaustive examination of the nature and conditions of the work; and such a course does not merely
throw discredit upon the undertaking itself, but also hinders
the execution of other similar enterprises founded upon
a sounder basis. In all very extensive works, unforeseen
contingencies are almost certain to arise, which may materially
modify the most careful estimates; and it is therefore specially
incumbent upon the designers of such works, to take every
possible precaution to reduce the chances of such occurrences
to a minimum. The execution of many large works would
no doubt have been delayed, if not prevented, if their ultimate
cost had been foreseen before the works were commenced;
but this affords no excuse for neglecting to make the most
careful investigations, in order to ascertain as closely as
possible the probable cost before commencing the work,
especially where, as in ship-canals across isthmuses, the canal
is perfectly useless until it is completed, and where, after the
first enthusiasm on the starting of the scheme has passed

away, any additional capital is obtained with difficulty, and generally under very onerous conditions.

## NICARAGUA CANAL.

The only scheme which has hitherto been regarded as a practicable alternative to the Panama Canal, for providing a waterway across the isthmus, is the Nicaragua route which, from its being nearer to the United States, has naturally been preferred by American explorers (Plate 13, Fig. 11). The Nicaragua route would, indeed, shorten the distance by water between the ports on the Atlantic and Pacific coasts of North America, by about 450 miles more than the Panama Canal; and between the United States ports on the Gulf of Mexico and the Pacific ports of North America by about 750 miles more[1]. American engineers have, moreover, frequently declared that the Nicaragua route offers a better, more feasible, and cheaper solution of the problem of interoceanic communication than the Panama route; and a Government Commission of the United States, after examining the surveys made under its direction at different parts of the isthmus, reported in 1876 in favour of the Nicaragua route as preferable to any other.

**Line selected for the Nicaragua Canal.** Various lines have been proposed from time to time for the construction of a canal across the isthmus within the State of Nicaragua, in all of which Lake Nicaragua, 102 miles long, about 42 miles in width at its broadest part, and of ample depth except near the shore, has formed the summit-level, with its highest water-level 110 feet above mean sea-level. A concession was granted by the Nicaraguan Government many years ago to an American Company for the construction of the canal, but no works appear to have been carried out; and a second concession was granted to the Nicaragua Canal Construction Company of New York in 1887. This Company obtained

[1] 'Mapas de Colton, América Central,' New York, 1889.

a charter of incorporation from the United States early in 1889; and preliminary operations were commenced the same year.

The line finally chosen for the canal starts from the Pacific at Brito, at the mouth of the Rio Grande; and owing to the steep slope on this side, the first lock is to be placed about 1½ miles only from the coast; and two more locks, about 3½ miles from the coast, will raise the canal to the summit-level, where the waters of the Rio Grande, its tributary the Tola, and other affluents, will be kept up by a dam across the valley at La Flor, forming the Tola basin (Plate 13, Figs. 9 and 10)[1]. From thence the canal will pass in cutting through the Western Divide, with a maximum depth of about 90 feet, into Lake Nicaragua, and after traversing the lake for 56½ miles, will utilize the river San Juan where it emerges from the lake which it drains.   The canal will then follow the course of the river San Juan for 64½ miles as far as Ochoa, where a dam will raise the level of this river, and its tributary the San Carlos, 56 feet, or nearly to the level of the lake, to provide a sufficient depth to allow of free navigation, and to reduce the amount of cutting beyond.   The canal will then diverge from the San Juan valley, and entering the San Francisco valley, where a lake will be formed by the high water-level retained, will pass on to the Eastern Divide, where the depth of cutting, even with the high water-level, will reach a maximum of 328 feet to the bottom of the canal. On emerging from the Eastern Divide, the canal will pass through the Deseado valley, forming another lake; and the summit-level will at length terminate at the eastern end of the Deseado lake, after a course of 154½ miles, where the first lock on the Atlantic slope is designed to be placed.   In the next 3½ miles, the canal will reach the level of the Atlantic Ocean through two additional locks; and after a course of about 12 miles beyond the last lock, it will

[1] 'Mapas de Colton, Trazado del Canal de Nicaragua,' New York, 1889.

emerge into the Atlantic Ocean at San Juan del Norte or Greytown.

The total length of the waterway by this route will be 169½ miles, of which only 26⅖ will consist of actual canal, the remaining 142$\frac{7}{10}$ miles providing free navigation through lake, basis, and river [1].

**Dimensions of proposed Nicaragua Canal and Locks.** In traversing soft soil, the canal is designed to have a bottom width of 120 feet, a depth of 28 to 30 feet and side slopes of 2 to 1 between Brito Harbour and the first lock, 3 to 1 between the last lock and Greytown, and 1½ to 1 elsewhere [2]. Through rock, the canal is to have a width of 80 feet, to be eventually increased to 100 feet, with vertical sides from the bottom up to about 4 feet above the water-level, where there is to be a bench 5 feet wide on each side, above which the rock is to be given a slope of 1 in 5; and the depth of the canal is to be 30 feet (Plate 11, Figs. 30 and 31). Where the river San Juan is not deep enough, in spite of the raising of the water-level by the Ochoa dam, it is to be dredged to a depth of 28 feet for a width of 125 feet; and the foreshores at the entrance into the lake at each side are to be dredged to a depth of 30 feet for a width of 150 feet.

All the six locks are to have an available length of 650 feet and a width of 80 feet. The lock nearest to Brito Harbour is designed to have a lift varying between 21 feet and 29 feet, according to the tides; and the two other locks on the Pacific slope are to have lifts of 42½ feet. The three locks on the Atlantic slope, descending towards Greytown, are designed to have lifts of 45 feet, 30 feet, and 31 feet respectively. The level of Lake Nicaragua ranges between 105 and 110 feet above sea-level, according to the volume of water discharged by the

[1] 'The Nicaragua Canal,' A. G. Menocal, Manchester Inland Navigation Congress, 1890, p. 4; and ' The Nicaragua Canal,' A. G. Menocal, Water Commerce Congress, Chicago, 1893, p. 25, table.

[2] ' Nicaragua Canal, 1893,' World's Colombian Exposition, Chicago, sections.

rivers draining into it; and allowance has been made for a slope of ¾ inch per mile in the surface of the river San Juan, from the lake to the Ochoa dam, for the discharge from the lake down the river when the lake is at its highest level; and consequently the total lift of the locks on the Atlantic slope has been made 4 feet less than of those on the Pacific slope.

**Excavation and Dams for the Nicaragua Canal.** The materials which the excavations for the canal will traverse, have been ascertained by borings to consist of clay overlying solid volcanic rock, in the deep cuttings through the two main ridges, and down the Pacific slope, clay predominating in the Rio Grande valley, and rock through the ridges. Rock is also met with on the western foreshore of Lake Nicaragua, and at the Toro and Castillo rapids on the river San Juan, where it will have to be blasted under water to secure the requisite depth; and mud, clay, and boulders will have to be dredged along 24 miles of the river above the Toro rapids. Stiff clay, with occasional boulders, extends from the east side of the Eastern Divide to the edge of the swampy stretch of land, interspersed with lagoons, bordering the Atlantic, 2 miles beyond the lowest lock; and from thence, sandy clay and sand are found as far as the coast. Sand with some mud and clay spreads over Brito Harbour; whilst only clean, sharp sand is met with in Greytown Harbour[1].

The dam at La Flor, for forming the Tola basin, is designed to be constructed of a central wall of concrete made with the rock excavated from the Western Divide, and an embankment of earth on each side. The dam will have a length of 2,000 feet; and the central wall will have a maximum depth of 170 feet for 1,000 feet of this length. It is proposed to form the dam across the river San Juan at Ochoa by tipping large blocks of stone across the river, and filling in the inter-

---

[1] 'The Nicaragua Canal,' A. G. Menocal, Water Commerce Congress, Chicago, 1893, p. 26.

stices with small materials; and the dam is to be given a slope of 1 to 1 on the up-stream side, and from 4 to 5 to 1 on the down-stream face. Embankments will also have to be formed across the side valleys opening into the San Francisco and Deseado valleys, to retain the water, against some of which there will be a head of water of over 60 feet, and 45 feet respectively.

**Discharge of Surplus Water from the Nicaragua Canal.** As the discharge into the canal from the rivers Grande, San Francisco, Deseado, and especially the San Juan and San Carlos, is considerably more at times than could be required for locking, outlets have had to be designed for the discharge of the surplus water, which would otherwise flood the canal. Weirs, accordingly, are to be provided on the right bank of the river San Carlos, with a total length of 1,200 feet, having their crests 2½ feet below the normal water-level of the canal at that part; and the crest of the Ochoa dam, 1,250 feet long, is to be placed 1 foot below the same level. Weirs also at the sides of the San Francisco and Deseado basins, with total lengths of 600 feet and 300 feet respectively, are intended to provide for the surplus discharge of these rivers; and a weir, 300 feet long, will pass the overflow from the Tola basin into the Rio Grande below.

**Harbours at the Entrances to the Nicaragua Canal.** A creek encumbered with sand and silt, through which the Rio Grande flows into the Pacific, situated in a slight indentation of the coast between two headlands, furnishes the only natural shelter at Brito. A harbour therefore is proposed to be formed by the construction of two converging breakwaters, projecting from the coast on each side of the creek, 1,000 feet and 850 feet long respectively; and the shallow portions of this sheltered area will be deepened by dredging.

Greytown, the only port on the Atlantic side for some distance along that flat swampy coast, is sheltered from the sea by a projecting spit of sand coming round from the south, the fringe apparently of the delta of the river San Juan, which since the diversion of the main discharge of the river from Greytown into the Colorado branch to the south, has gradually been closing up the port under the influence of the waves, converting it at last into a lagoon about the year 1860. To open up this harbour again, it has been proposed to carry out a rubble mound breakwater 3,000 feet from the shore, extending into a depth of 6 fathoms. This breakwater is intended to act like a groyne in arresting the progress of the sandy spit; and an outlet channel would be dredged under its shelter, connecting the lagoon again with the sea.

**Estimated Period of Construction, and Cost of the Nicaragua Canal.** The Ochoa and Tola dams, the harbours, and locks have been considered as likely to occupy four to five years in construction; but six years have been allotted to the great cutting through the Eastern Divide, which has consequently been taken as the period required for the completion of the canal.

The total cost of the canal works, including 25 per cent. for contingencies, has been estimated at about £13,500,000 [1]; but making further allowance for unforeseen contingencies, and adding interest during construction, commissions, and other charges, the total cost of the undertaking has been fixed at £20,800,000 [2].

**Commencement and Stoppage of the Nicaragua Canal Works.** Surveys of the site of the canal were carried out in 1888; preliminary works were commenced in 1889; and in 1890, the rivers San Juanillo and Deseado were cleared out for a distance of 30 miles, to enable stores to be transported along them. The line of the canal also was cleared for 10 miles from Greytown, and for 9 miles on the Pacific slope; and

---

[1] 'The Nicaragua Canal,' A. G. Menocal, Water Commerce Congress, Chicago, 1893, p. 38.

[2] 'Nicaragua Canal, 1893,' World's Columbian Exposition, p. 12.

a railway was commenced towards the Eastern Divide. By 1893, railways were in course of construction between Greytown and the Eastern Divide, 12 miles of which were in operation, and between Brito and the lake; and 1,000 feet of a provisional timber jetty filled with stones had been carried out at Greytown, to obtain a sufficient depth of channel at the entrance for landing plant and stores. The Company, however, after expending about £800,000 on these works, was unable to procure funds to meet its outstanding liabilities, so that in the middle of 1893 a receiver was appointed, and the works suspended. There appears to be little prospect of a resumption of the works, unless the enterprise is taken over by the United States; and a bill was introduced for this purpose at the close of 1894, but failed to receive the assent of the legislature. A Commission, however, has been appointed by the United States Government to investigate and report upon the feasibility and cost of constructing the canal [1]; and upon this report will probably depend the future of this undertaking.

**Remarks on the Nicaragua Canal.** The scheme possesses the advantages of providing a considerable length of free navigation through Lake Nicaragua and in flooded valleys, an abundance of water for locking stored up by the lake forming its summit-level, and a moderate elevation to be surmounted by three locks of large lift on each slope. It has been estimated that vessels will be able to traverse the waterway in 28 hours, at a speed varying, according to the expanse of water, from 5 miles an hour in the canal proper, up to 10 miles an hour in the open lake, 45 minutes being allowed for the passage of each lock, and one hour for detentions.

The works, however, as proposed are open to some serious objections. The method of construction proposed for the

---

[1] 'The Times,' 'The Nicaragua Canal,' October 28, 1895. The report appears to have been drawn up, but has not yet been presented to Congress. Some extracts, however, just published appear to indicate that the report is unfavourable both as regards construction and cost.

Ochoa dam does not appear to be solid enough to withstand high floods passing over its crest, especially considering the uncertain and shifting foundation of the river-bed ; and this dam, the high reservoir dam of the Tola basin, and the high embankments for retaining the water along the San Francisco and Deseado valleys, would be in danger of dislocation, and resulting destruction, in the event of the occurrence of a serious earthquake in a district adjoining numerous active volcanoes. If the fact that 'an earthquake might cause serious damage to a masonry dam' is an adequate reason for not adopting a masonry dam at Ochoa, the La Flor dam, with its proposed high central wall of concrete, could not be regarded as secure ; whilst the suggestion that an earthquake could do no harm to the Ochoa dam as proposed, but, 'on the contrary, it may add to its consolidation by bringing the parts in closer contact'[1] does not tend to promote confidence in the soundness of the engineer's judgment. The maintenance, moreover, of deep-water access at Greytown may prove a serious difficulty on an advancing foreshore encumbered by sands brought down by the rivers ; and this harbour seems likely to require a large expenditure in its formation and maintenance, and to tax the highest skill of the engineer. The great cutting, also, through the Eastern Divide, may prove a much more difficult task than anticipated by the promoters ; for though a more thorough examination of the strata of this ridge appears to have been made than of the Culebra ridge at Panama, it is very difficult to determine by borings the nature of the clay and compactness of the rock, and to what extent these strata may be disintegrated when exposed to tropical rains. The rainfall, indeed, on the Atlantic slope of the Nicaragua Canal appears to be quite as heavy as at Panama, having averaged 267·45 inches at Greytown in the three years 1890–92, the recorded maximum reaching 6 inches in 24 hours ; whilst, though the

---

[1] 'The Nicaragua Canal,' A. G. Menocal, Water Commerce Congress, Chicago, 1893, p. 30.

lake region and the Pacific slope may be comparatively fairly
healthy, the swampy land bordering the Atlantic and the
adjoining country appear to be extremely unhealthy. The
control of the floods of the rivers discharging into the waterway on the Atlantic slope, may prove somewhat difficult;
whilst it is very probable that considerable quantities of the
sand and silt brought down by the river San Carlos will find
their way into the canal.

Under all these circumstances, and considering the magnitude of the earthworks, the great care needed in the construction of the dams and embankments liable to be disturbed
by earthquakes, and the uncertainties attending the formation
and maintenance of the harbour at Greytown, it appears that
the estimate put forward is inadequate for the works, and that,
like at Suez and Corinth, a largely increased expenditure will
be needed before the canal can be completed. A comparison,
indeed, of the longitudinal sections of the Suez and Nicaragua
Canals (Plate 13, Figs. 2 and 10), combined with a consideration
of the differences in the rainfall, climate, and wages at the two
sites, points to the conclusion that the Nicaragua Canal
involves much more difficult problems, and necessitates a
larger amount of work, than the Suez Canal did, and that
it will inevitably cost a considerably larger amount in
construction.

## COMPARISON OF PANAMA AND NICARAGUA CANALS.

Panama and Nicaragua appear to be the only two sites at
which the construction of a canal across the isthmus appears
at all reasonably feasible; and it may be anticipated that
sooner or later a waterway will be completed along one
of these routes, connecting the Atlantic and the Pacific, with
great benefit to navigation between Europe and Eastern
Asia, and more especially between the eastern and western
coasts of North America and the northern portion of South
America.

**Contrast between the Panama and Nicaragua Routes.**
The two routes, only 280 miles apart on the Atlantic side, present a somewhat remarkable contrast, though both cross the isthmus at places where there is a considerable depression in the chain of the Cordilleras (Plate 13, Figs. 5, 6, 9, and 10). The Panama route is only a little more than a fourth of the length of the Nicaragua route; but it traverses a single central ridge about 11 miles across, and two short rises on the Atlantic slope at Bohio-Soldado and San Pablo, and was designed to be of a narrow section throughout, except at the passing-places. The Nicaragua route, on the contrary, though much longer, has, like the Suez Canal, the advantage of possessing a long length of lake navigation; and this free navigation will be further increased by damming up the rivers, without which, indeed, the formation of a waterway on the Atlantic slope at that site would be impracticable. In spite, however, of this raising of the water-level, there will be a large amount of excavation for the Nicaragua Canal, under conditions at the great cutting of the Eastern Divide, apparently resembling those at Panama, though the strata at Nicaragua may possibly be more favourable.

**Problems involved in forming a Waterway across the Isthmus of Panama.** The great problems to be investigated for the completion of the Panama Canal with locks, are the possibility of controlling the river Chagres, and the excavation of the deep cutting through the Culebra ridge, with its treacherous soil, though greatly reduced in depth. The great difficulties at Nicaragua, which do not appear hitherto to have received due consideration, are the high dams and embankments required in a region subject to earthquakes, the physical obstacles to providing deep-water access at Greytown harbour, and the very deep cutting at the Eastern Divide, exposed to an excessive rainfall and in a very unhealthy climate.

**Prospects of the two Schemes for a Waterway across the Isthmus of Panama.** A very complete investigation of the two sites could alone enable a determination to be made as to which scheme, in its present condition, is the most feasible and least costly. The higher summit-level adopted at Panama than the level of Lake Nicaragua, involves two additional locks on each slope; and, except through rock, the section of the Nicaragua Canal is designed to be considerably wider than that adopted for the Panama Canal; whilst the cuttings at Nicaragua have the advantage of being broken up into sections, mostly of moderate length, with a maximum length of about 9½ miles through the Western Divide, instead of the continuous cutting at Panama. One scheme, however, for the completion of the Panama Canal introduces the lake system with dams for one or two of the reaches on the Atlantic slope, which, besides possibly affording a means of dealing with the floods of the Chagres, would also provide lake navigation for several miles in place of a narrow canal. Panama is some distance from the region of active volcanoes, and therefore dams there would be less exposed to disturbance than at Nicaragua; whilst the great reduction in the depth of the main cutting, by the introduction of locks, would greatly facilitate a widening of the canal. Judging, indeed, by the longitudinal sections, and considering the excavation already accomplished at Panama, it seems probable that the earthwork required for completing the Panama Canal with locks, even with a wider section, would be less costly than for the formation of the Nicaragua Canal. Accordingly, if the completion of one of these waterways depended merely upon a dispassionate choice of the most feasible of the two schemes, it is quite possible that, in spite of the discredit attaching to the late Panama Canal Company, and the large sums wasted upon the undertaking, the balance of advantages might be found to be in favour of the Panama route, provided the floods of the Chagres could be dealt with at a reasonable cost. If, moreover, the prospects of raising the necessary capital depended solely upon the French and American nations, it seems probable that, after the lapse of time shall have somewhat obliterated the memory of the Panama Canal scandals, the funds for completing the Panama Canal would be more readily obtained in France, where the scheme is regarded as a national enterprise, than the requisite capital subscribed in the United States for the construction of the Nicaragua Canal, which has never been regarded as in any sense bound up with the national honour, as amply shown by the comparatively small amount raised for carrying out the scheme.

When, however, the assistance of a Government is invoked, which may prove the only method of carrying out such a vast undertaking, and without which the Suez Canal could hardly have been accomplished, the conditions are essentially altered. The French Government has no more special interest in the construction of a waterway across the Isthmus of Panama than the rest of the maritime countries of Europe, beyond the consideration that it would be a popular measure with a section of the French nation; and probably, as in the case of the Suez Canal, British vessels would provide the bulk of the international traffic through the canal. Moreover, the intervention of a European Government in a matter that might invest it with special rights of a novel and important character in Central America, would introduce international complications. The Governments of Colombia and Nicaragua do not possess the resources for such an undertaking, nor could they be entrusted with the control of such a waterway. The United States is the only Government that could, under the circumstances, undertake the construction of a waterway across the Isthmus of Panama, and its subsequent control, safeguarded by international agreements. The Nicaragua Canal, owing to its position, is so distinctly the best for the United States, that, unless further investigations should prove that the scheme is impracticable at any reasonable cost, or that there is serious risk of the proposed dams and embankments being dislocated by earthquakes, this route is certain to be adopted by the United States Government if it takes the matter in hand, even if the estimated cost of the construction of this waterway should exceed that of the completion of the Panama Canal. Under these conditions, there is a better prospect of the construction of the Nicaragua Canal than of the completion of the Panama Canal, the works of which in the first instance were far more vigorously prosecuted.

## VARIOUS SCHEMES FOR SHIP-CANALS.

No schemes for ship-canals remain to be entered upon, for shortening the routes of ocean-going traffic, at all equal in interest or world-wide importance to the Suez Canal and the waterways commenced across the Isthmus of Panama. Several schemes, however, have been proposed for cutting across comparatively narrow necks of land, which would materially shorten the present sea routes, and for converting important inland cities into seaports.

**Isthmian Ship-Canal Schemes.** In Europe, two schemes for constructing a ship-canal across Scotland between the Forth and the Clyde, suitable for the largest class of vessels, $69\frac{1}{4}$ miles, and 30 miles long respectively, have been designed for the purpose of saving a stormy, circuitous journey round the North of Scotland, and also to give Glasgow an outlet to the North Sea [1].

In Asia, the Manaar Canal which it is proposed to cut through the reef extending across the channel between India and Ceylon, at the island of Rameswarran, would connect the Gulf of Manaar with the Palk Straits by a navigable

---

[1] 'Proposed Forth and Clyde Ship-Canal,' D. A. Stevenson; and 'Proposed Forth and Clyde Direct Route Ship-Canal,' J. Law Crawford, Manchester Inland Navigation Congress, 1890.

channel, and thus reduce the voyage between Colombo and the Bay of Bengal by 350 miles. Various projects have been brought forward for forming a navigable waterway across the Isthmus of Kraw which connects the Malay Peninsula with Siam, and has a minimum width of 44 miles. By connecting the river Pakshan on the west slope with the river Champon on the east slope, by a canal traversing the central ridge rising to a height of 250 feet, but only 7½ miles in width, the Bay of Bengal would be joined to the Bay of Siam, thereby shortening the voyage between Ceylon and Hong Kong by 300 miles, and between Calcutta and Hong Kong by 540 miles.

Africa, from its great breadth, is unsuited for the formation of ship-canals across it, with the single exception of the Isthmus of Suez ; but a proposal has been made to cut a canal from the Mediterranean to the desert in the interior, for flooding some vast depressions to the south of Algeria and Tunis. An inland sea, 3,160 square miles in extent, would thus be formed, which it is anticipated would greatly improve the climate of the surrounding district, and would render the adjacent lands capable of cultivation ; whilst the canal would also provide easy means of access to the sea[1].

In North America, the construction of a ship-canal has been proposed, traversing the northern part of the peninsula of Florida, about 158 miles in length, starting from Cumberland Sound on the Atlantic Coast, and entering the Gulf of Mexico at the mouth of the St. Mark's River[2]. This canal would shorten the voyage between the North American ports on the Atlantic and the mouth of the Mississippi and Galveston harbour by about 460 miles, and 430 miles respectively.

The great difficulties attending all schemes for large ship-

---

[1] 'Mémoires de la Société des Ingénieurs Civils,' 1883, (2) p. 484, and plates 64 and 65.

[2] 'Map showing the Location of Works and Surveys for River and Harbour Improvement,' Washington, 1879.

canals consist in the unforeseen contingencies to which such extensive works are subject, and the impossibility of estimating with any degree of accuracy the amount of traffic which will make use of the new route.

Ship-Canals proposed for Inland Cities. The authorities of some large inland cities have of late years evinced a strong desire to secure the same advantages of sea-going trade as those enjoyed by towns on the sea-coast or on large tidal rivers  Thus for several years, schemes have been urged for the conversion of Paris into a seaport, though situated 70 miles inland from Rouen in a straight line, in a recent one of which a tidal channel, 19⅔ feet deep, was proposed to be excavated along the valley of the Seine up to Poissy, 12½ miles from Paris, where a port was to be created, and from which vessels could reach Paris by a chain of locks [1]. A subsequent modification of the scheme provides two intermediate locks at Poses and Méricourt, with lifts of 20⅔ feet, and 22 feet respectively ; and a lock at Poissy, and another at Sartrouville, would bring the vessels up to Paris, where the water-level would be 73¼ feet above low water at Rouen [2]. The length of this ship-canal, following for the most part the course of the Seine, would be 115 miles in place of 150 miles by the river, with a depth of 19⅔ feet instead of 10½ feet as at present ; and the estimated cost is £6,000,000. The Government has, however, provided an excellent inland waterway along the Seine between Rouen and Paris; the proposed ship-canal would involve a very large expenditure, and would very seriously modify the conditions of the valley of the Seine; and Rouen appears the natural terminus of ocean-going traffic along the Seine.

Brussels, which at present is connected with the Scheldt

---

[1] 'Résumé d'une Étude sur la création d'un Port de Mer à Paris,' A. Bouquet de la Grye, Paris, 1882 ; and Association Française pour l'Avancement des Sciences, Rouen, 1883, pp. 227 and 230.

[2] 'Paris Port de Mer,' A. Bouquet de la Grye, Paris, 1892, p. 122, and plate 2.

by the river Rupel up to Willebroeck, and from thence by a canal 16 miles long, with a depth of 10½ feet and accessible to vessels of 300 to 400 tons, is to be transformed into a seaport by the enlargement of the canal, and by increasing its depth to 19⅔ feet, so as to enable vessels of 2,000 tons to reach Brussels. Three new large locks are to replace the three existing locks on the canal. The cost of this ship-canal, together with the formation of a port between Laeken and Brussels, is estimated at £1,400,000, three-fifths of which sum is to be provided by the city and the neighbouring communes, and the remainder by the State and the provinces.

A scheme is under consideration for converting Berlin into a seaport, by the construction of a ship-canal at an estimated cost of £10,000,000. The canal is designed to have a width of 70 feet at the bottom and 190 feet at the water-level, and a depth of 25 feet. The canal, as proposed, is to proceed northwards from Berlin towards the Finow Canal, which it is intended to utilize for a short distance up to Hohensaaten, from whence the canal is to follow the valley of the Oder down to Greifenhagen, where it is to join the river, and thereby obtain an outlet into the Baltic at Swinemünde. The fall from Berlin to the Oder would be provided for by three locks; and vessels passing down the canal would cross the Baltic to Kiel, and reach the North Sea through the Baltic Canal.

**Concluding Remarks.** The above references to schemes for ship-canals show that ample opportunities still exist for the improvement of the routes of ocean-going trade, by the construction of ship-canals across isthmuses, and for the conversion of inland cities into seaports by similar works. Schemes, however, for ship-canals across isthmuses imperatively demand the most careful investigation of the site before starting such a work; and ample allowance should be made in the estimates for the contingencies which so often arise during the progress of these works. Nevertheless, in spite of the most careful prevision, works of such magnitude and extent must always be exposed to the difficulties of raising a large capital on reasonable terms, the occurrence of unforeseen contingencies, and the uncertainty attending the estimates of the probable traffic through the canal.

Ship-canals to large cities, like those to Berlin and Brussels, must depend largely for their financial success upon the physical conditions of the route, and the commercial position of the places to which they give access. Such schemes, however, may be more advantageously carried out by the State than by a private Company, for the State not only can raise the large sums required upon much easier terms, but it also profits by the indirect advantages resulting from the growth of its principal cities and the expansion of trade ; whereas the returns of a private Company are restricted to the earnings of the canal. Where State aid is not available for such undertakings, and the country possesses an extensive sea-coast like Great Britain, it is wiser not to attempt to carry out the extensive works necessary to bring sea-going vessels up to towns at some distance from the sea, but to rest content with the formation of enlarged inland canals suitable for vessels of 300 to 600 tons, like the schemes proposed for Birmingham and Sheffield ; and such canals, being available for sea-going vessels of moderate size trading with the ports of neighbouring countries, afford some of the facilities provided at seaports.

# INDEX.

A.

of, 445; area irrigated by, 445; oblique weir across Durance for directing flow towards head of, 445; cost of, per acre irrigated, 459.

Amazon River, basin and length of, 4; appearance and height of bore on, 242; natural highway for inland navigation, 469.

Amsterdam and Merwede Canal, time of completion and cost, 489; description of, 489; dimensions of locks on, 489; size of largest vessels admitted into, 489; in communication with Amsterdam Ship-Canal, 566.

Amsterdam Ship-Canal, 563–571; sand-pump dredgers used on, 85; cost of dredgers and dredging at, 85; floating tubes for deposit of dredgings, 94, 355, 565; wooden and iron lock-gates for, 104, 567; dimensions of, 351; locks of different sizes with intermediate gates on, 375, 567; dimensions of large North Sea lock on, and of new lock, 377, 568; proposed at various times, 563; concession granted for construction of, 563; shortening route to North Sea, 563; obstacles to, and arrangements adopted for, 563–564; rise of tide at each end, 565; description of, 565–566; amount of excavation for, 566; waterways communicating with, 566; swing-bridges across, 566; branch canals from, to villages bordering Lake Y, 566; period of construction of, 566; embankment shutting off Zuider Zee from, 566; Zuider Zee locks on, 566–567; pumping machinery for discharging drainage waters from, 567; volume of water discharged from, 567–568; locks on, at Ymuiden, 568; estimated cost of new lock for, 568; harbour at North Sea end of, 568–569; enlargement of, 569–570; largest vessels hitherto admitted into, 570; largest vessels which could pass through new lock, 570; cost of, 570; receipts from sale of lands reclaimed by, 570; traffic on, 570–571; compared with Baltic Canal, 579–580.

Amur River, large river of Asia, 4.

Anderton Canal Lift, 401–404; connecting the Weaver with Trent and Mersey Canal, 401; description of, 401–403; method of working, 403; height of lift and time occupied, 403; saving of time and water as compared with flight of locks, 403; cost of, and of foundations for, 404; accident to, 404.

Anicuts, 111–112; object of, 111; temporary, in former times, 111; permanent construction of, 112; Okhla, 112; Godavery, 112; Dehree, 112; sluices formed in, 112.

Aquamotrice, form of bag and spoon dredger, 72–73; description of, 72–73; amount and cost of material raised by, on Garonne, 73.

Aqueducts for Canals, 360–361; carrying canals over roads, railways, and valleys, 360; points of difference from ordinary bridges, 360; constructed at Briare to do away with transit of Loire, 360–361; old, at Barton, 361, replaced by swing, across Manchester Ship-Canal, 361, 589; across Solani for Ganges Canal, 439; across Kali Nadi for Lower Ganges Canal, 440; Roquefavour, for Marseilles Canal, 446; across Paronvier Valley for Verdon Canal, 447; siphons in place of, on Verdon Canal, 447–448; on Carpentras Canal, 450; on Henares Canal, 451; timber flume serving for, on Highline Canal, 453; carrying Bear River Canal across Malad Valley, 454; conveying Turlock Canal over valleys, 456.

Aragon Canal, period of construction, 451; volume of water supplied to, by Ebro, 451; length of, and area irrigated by, 451; cost of, per acre irrigated, 451; rendered navigable, 497; size of vessels navigating, 497; traffic diverted from, by railway, 497.

Arizona Canal, 453–454; supply for, from Salt River, 453; crib-work weir with rubble across Salt River for supply of, 453; headworks of, 453–454; description of, 454; area irrigated by, 454; cost of, per acre irrigated, 459.

oceanic canal still in existence, 597;
used for inland navigation, 597; advantages of proposed enlargement of,
597; unfavourable physical conditions
for forming, into ship-canal, 597;
difficulties and vast cost of scheme,
598.

Canal du Centre, connecting basins of
Saône and Loire, 363, 481; supplied
by reservoirs retained by earthen dams,
366; reduction in number and increase
in lift of locks on, 379–380; cylindrical
sluice-gates and locks on, 385; period
of opening, and extent of waterway
completed, 481; traffic on, 485.

Canal du Centre of Belgium, to connect Mons and Condé Canal with
Charleroi and Brussels Canal, 408,
492; length and rise of, 408, 492;
La Louvière canal lift on, 408–410;
depth of, and size of locks and lifts
on, 493.

Canal Inclines, 389–398. See Inclines
on Canals.

Canal Lifts, 398–413; contrasted with
canal inclines, 398; primitive, on
Worcester and Birmingham Canal,
398–399; on Grand Western Canal,
399–400; hydraulic, at Anderton on
the Weaver, 401–404; at Fontinettes
on Neuffossé Canal, 404–408; La
Louvière, on Canal du Centre of
Belgium, 408–410; remarks on, in
connection with inclines and locks,
410–414.

Canalization of Rivers, 64–70; object
and advantages of, 64, 67; introduction of stanches, 64–66; effected by
locks and weirs, 66–67; Seine and
Main examples of, 67–69; remarks
on, 69–70.

Canals, instead of delta channels, 198–
200; definition of, 343; for conveyance of water, or for navigation, 343–
344; drainage, 344; irrigation, 344–
346; navigation, 346–347; connecting
rivers, 347–348; lateral to rivers, 348–
351; sizes of, for navigation, 351–353;
construction of, for navigation, 353–
354; earthwork for, 354–356; bridges
over, 356–360; aqueducts for, 360–

361; tunnels for, 362–364; waste
weirs and sluices for, 364; stop-gates
on, 364–365; supply of water for, 365;
reservoirs for supplying, 365–370;
introduction of supply to, 370–371;
consumption of water on, 371–372;
remarks on works for, 372–373; contrast of works for, with those for railways, 372–373; conditions necessary
for competition of, with railways, 373;
locks for, 374–375; sizes of locks for,
375–378; lifts and flights of locks for,
378–382; reduction of time in locking
on, 382–385; methods of saving water
in locking on, 385–387; remarks on
locks for, 387–388; inclines and lifts
in place of locks for, 389; inclines on,
and methods of conveyance along,
389–398; lifts on, 398–410; remarks
about inclines and lifts on, 410–414;
classification of irrigation, 425; irrigation, supplied from reservoirs, 425,
426; inundation, 426–431;

perennial irrigation, — definition,
432; classification, 433; upper, 433–
460; deltaic, 460–465;

conditions affecting construction
of, for inland navigation, 469–470;
connecting rivers, 472–474; extension of, in France, 481–482; in
Holland, 486–488; in Belgium, 491–
493; Spanish, 497; navigable, in
Italy, 498–501; in Sweden, 503–504;
in Germany, 507–509; in Austria,
511; in Hungary, 512; in Russia,
516–518; navigable, in India, 520–
521; in Canada, 521–525; navigable,
in United States, 527–535; ship-,
550–651. See also Ship-Canals,
Irrigation Canals, Inland Navigation.

Canals for Navigation, described—
Amsterdam, 563–571; Amsterdam
and Merwede, 489; Baltic, 571–580;
Bruges, 593–595; Caledonian, 598–
599; Chicago Drainage, 531–533;
Corinth, 615–621; Ghent-Terneuzen,
554–556; Gloucester and Berkeley,
551–552; Loire, 308, 552–553; Manchester, 580–593; Marie, 516–517;
Nicaragua, 635–647, North Holland,

561–563; Panama, 625–635, 643–647; Perekop, 621–622; St. Louis, 556–557; St. Mary's Falls, 376–377, 530; Sault-Sainte-Marie, 522, 524; St. Petersburg and Cronstadt, 557–559; Suez, 600–615; Tancarville, 317–318, 553–554; Welland, 376, 522, 524. See Amsterdam Canal, &c.; and see also Irrigation Canals.

Canals for Navigation referred to:— Agra, 520; Aire and Calder Navigation, 376, 471; Aragon, 497; Arles and Bouc, 556;

Beauharnois, 522, 524; Bega, 512; Berizinski, 517; Birmingham, 366, 476; Bourgogne, 473, 482; Briare, 361, 481; Bridgewater, 361, 475–476; Brussels and Rupel, 493, 494; Buckingham, 520, 521;

Caer Dyke, 475; Calcutta and Eastern, 520–521; Canal de l'Est, 348, 483; Canal des Deux Mers or Canal du Midi, 481, 597–598; Canal du Centre, 379–380, 481; Canal du Centre, Belgium, 408, 493; Castile, 497; Chambly, 522, 525; Champlain, 528, 534; Chard, 394–395; Charleroi and Brussels, 493, 494; Chesapeake and Ohio, 528, 534; Corsini, 499, 501; Crinan, 366;

Dalsland, 504, 505; Dnieper-Bug, 367, 517; Don and Volga (proposed), 348, 518; Duke Alexander of Wurtemburg, 518;

Eems, 487; Eider, 571; Ellesmere, 366; Ems-Jade, 507; Erie, 527–528, 533; Eskilstuna, 503; Exeter, 376;

Finow, 507; Fiumicino, 498, 499; Forth and Clyde, 474, 477; Foss Dyke, 475; Franz, 512;

Ganges, 436, 520; Garonne Lateral, 349, 485; Ghent and Ostend, 492, 493; Glasgow and Paisley, 351; Godavery, 520; Göta Navigation, 504, 505; Grand Junction, 476, 478; Grand Western (Tiverton), 399–401; Hennepin, 531, 535; Huddersfield, 362;

Illinois and Michigan, 528, 534; Kennet and Avon, 348; Kinda, 504, 505; Kistna, 520; Kurnool, 520;

Lachine, 522, 524; Languedoc, or Canal du Midi, 481; Leeds and Liverpool, 378, 476; Loing, 348, 485; Loire Lateral, 349, 485; Lower Ganges, 520; Louisville and Portland, 527, 534;

Mahmoudieh, 199; Main-Danube, or Ludwigs, 474, 507; Marne Lateral, 349, 483; Marne-Rhine, 472–473; Marne-Saône, 367, 483; Midnapore, 520; Monkland, 351, 395; Mons and Condé, 408, 495; Morris, 392, 528; Murray, 523; Muscle Shoals, 530;

Naviglio Grande, 498, 501; Neuffossé, 404, 549; Nivernais, 482;

Oberland, 393; Obi-Yenisei, 472; Oder-Spree, 507; Oguinski, 517; Oise-Aisne, 483; Orissa, 520; Orissa Coast, 520, 521; Ostia, 198–199; Oswego, 208; Ottawa, 523, 524; Ourcq, 359, 393;

Pavia, 498, 500; Pontine, 498, 499; Rhone-Rhine, 473, 509; Rideau, 523, 524; Rochdale, 366, 476; Rupel and Louvain, 493;

Saar, 507, 509; St. Clair Flats, 530–531; St. Denis, 380–382; St. Lawrence, 376, 522; St. Quentin, 481, 485; Sambre-Oise, 482–484; San Carlos, 199, 496; Scheldt-Meuse, 491, 492; Shrewsbury, 391; Shropshire, 391; Sirhind, 520; Södertelge, 504, 505; Somersetshire, 378; Sone, 520; Soulanges, 524; South Beveland, 487, 490; South Willems, 487, 490;

Thames and Severn, 363, 474; Trent and Mersey, 362, 476; Trent Navigation, Canada, 523; Trollhätta Navigation, 503–504;

Valle, 520; Voorne, 222, 487; Vychni-Volotchok, 516;

Walcheren, 487, 490; Warwick and Birmingham, 378, 478; Western Jumna, 520; Worcester and Birmingham, 378, 478;

Yypres-Lys, 492.

See Agra Canal, Aire and Calder Navigation, &c.; and see also Irrigation Canals.

161; advantages of low banks for, 161-162; objects and disadvantages of high embankments, 162-163; of Mississippi, 163-164; of Po through Lombardy, 164-165; of Theiss, 166; raising of river-bed produced by, 167-168; effect of, in upper valleys, 169; success of, for Upper Rhone, 169; advantages and disadvantages of, 170-171.

Ems River, portion of, canalized, 506; connected by canals with the Jade and Hunte, 507.

Ems-Jade Canal, connecting the Ems with the Jade, 507; depth of, 507.

Enlargement of Channel, rate of, 323-325; irregular, in untrained estuaries, 323; instances of, in rivers, 323-324; examples of, in trained channels, 324; limits of suitable rates for, 324.

Erie Canal, providing a waterway between New York and the lakes, 347; route and length of, 527-528; original and enlarged dimensions of, 533; proposed reconstruction of, 533-534; traffic on, 535.

Erne River, water-level of, regulated by weir at Belleek, 115-116.

Eroding Machines, 87-88; rakes and harrows tried on bars of Danube and Mississippi, 87; rakes used on Stour and Danube, 87; scrapers employed on Upper Mississippi and Missouri, 87; revolving cylinder with spikes drawn along Fen rivers, 87; description of 'eroder' used on Witham, 88.

Eskilstuna Canal, first canal in Sweden constructed with locks, 503; connecting Eskilstuna and Lake Hjelmar with Lake Mälar, 503.

Esla Canal, supply for, drawn from river Esla, 451; area irrigated by, 451; cost of, per acre irrigated, 459.

Esla River, valley of, irrigated by Esla Canal, 451; masonry weir across, for supply of Esla Canal, 451.

Estuaries, 245-249; instances of tidal, 18; contrast presented by tidal, to outlets of tideless rivers, 173; expediency of gradual enlargement of, towards outlet, 246; variety in forms of, 246; instances of irregular, 246; instances of, with contracted outlet, 247; instances of tidal rivers devoid of, 247; effect of form on condition, 247; remarks on influences of forms of, 248-249; training works in, 281-326; experimental investigations on training works in, 327-342; uncertainties as to effects of training works in, 327-328.

Eure River, ratio of discharge of, to rainfall, 13.

Evaporation, 8-9; causes of variation in, 8; effect of, on Caspian, Dead Sea, and Lake Chad, 8; influences producing maximum and minimum, 8-9; reduced by percolation, 9; loss from, in tanks in India, 420.

Excavators, 95-99; importance of, for ship-canals, 71; description of—bucket-ladder, 95-97; steam navvy, 97-98; grab, 98; remarks on, 98-99; used in large cuttings on ship-canals, 354-355; for forming Baltic Canal, 573; number of, used on Manchester Canal, 584; number of, employed for Panama Canal, 630.

Exeter Canal, dimensions of lock on, 376.

Experiments with Model Estuaries, 327-342; objects of, and methods of conducting, 327-331; of tidal Seine, and results, 331-337; of Mersey estuary, 337-342; conclusions from, 342; remarks on, 342.

F.

Fall of Rivers, 14-15; variation in, 14; diagram of, 15; definition of, and mode of expressing, 22; in Fen rivers, 154; loss of, by obstructions, 157; increased by straight cuts, 160; small, in case of Volga, 178; improvement of, in tidal portion of Fen rivers by straight cuts, 284-288.

Fascine Mattresses, for forming training banks, 54; in Mississippi South Pass jetties, 194-195; used for regulating head of Mississippi passes, 195 ; jetties at new mouth of Maas formed with, 222; in training works at outlet of new Witham outfall, 286; employed for training works of Weser, 301.

Fen Rivers, 284–289; small fall of, 154; catch-water drains used in basins of, 154; advantages of enlarging channel of, 160; value of straight cuts for, 160; pumping for draining low lands of basins of, 166; training works and straight cuts for improving outfalls of, 284–289.

Finow Canal, length of, 507; connecting the Havel with the Oder, 507; to be partially utilized for Berlin Ship-Canal, 650.

Fiumicino Canal, for avoiding delta of Tiber, 498; length and size of, 499; traffic on, 501.

Fixed Weirs, 108–112; form, and disadvantages of, 108; modifications of, 109; oblique, on Severn, 109; angular and horse-shoe, 110; on River Lot, 110; construction of, 111; anicuts in India, 111–112; advantages and defects of, 119; of masonry across Ticino for Naviglio Grande, 443; of fascines, timber, and boulders, across Durance for Crapponne and Alpines canals, 445; of masonry with movable shutters across Durance for Marseilles Canal, 446; of masonry across Verdon for Verdon Canal, 448; of timber, fascines, and boulders across Durance for St. Julien Canal, 449; of masonry across Henares and Esla for canals, 451; of crib-work and rubble across Platte for Highline Canal, 453; of crib-work across Salt River for Arizona Canal, 453; of crib-work for Bear River Canal, 454.

Flights of Locks, made double to reduce consumption of water, 371–372, 379; instances of, 378–379; advantages of, 379; replaced at Neuffossé Canal by lift, 404; on Caledonian Canal, 599.

Floating Tubes, 94–95; for discharge of dredgings on to banks—in Amster-

dam Canal, 94, 355, 568; in enlargement of Ghent-Terneuzen Canal, 94–95, 355; in construction of Baltic Canal, 573.

Floats, 25–27; surface, 25; objections to surface, 25; forms of double, 26–27; tube, 27; velocity-rods for, 27; views about double, 33; objections to double, 33; advantages of tube, and limited use of, 34; tidal ebb and flow on Clyde observed by, 255.

Flood and Ebb Tides, relative influences of, in a river, 236; advantages of uniformity in action of, in a river, 236; results of conflicting action of, 236–237; instances of blind channels formed by flood, 237.

Floods, 148–171; value of forests in reducing, 9; increased by clearing forests, 10; rarity of, in Seine in warm season, 10; influence of winter rains on, 10; greatest, produced by rain on melting snow, 10; summer, caused by melting of glaciers, 11; influence of strata on, 11–12; differences in time of arrival of, from tributaries, 12; value of movable weirs for passage of, 121–122; causes of, 148–149; prediction of, 149–152; conditions influencing height of, 149–150; in Seine basin, 150; arrangements for predicting, 151; value of predictions of, 151–152; methods of providing protection from, 152–171; caused by breaches in Mississippi levees, 163; in valley of Po, 164, 165; increased height of, on Po, Adige, and Reno, 165; in Loire valley, 166; of Theiss, 166; of Yellow River, 167–168; of embanked Japanese rivers, 168; rating for protection from, 170; measures for mitigating, 170–171; mitigation of, by improvement of river outlets, 171.

Florida Ship-Canal (proposed), route proposed for, 648; length and course of, 648; object of, 648.

Fontinettes Canal Lift, 404–408; to accommodate large traffic on Neuffossé Canal, 404; description of, 405; weight and height raised, 405–406;

## G.

Ganges Canal, 439; form of double
float used for gauging flow of, 27;
velocity-rods employed for obtaining
mean flow of, 27; correspondence of
gaugings on, with Mississippi formula,
46; adapted for navigation, 436, 520;
course of, 439; aqueduct conveying,
across river Solani, 439; discharge
of, 439; length of, and area irrrigated
by, 439; cost of, per acre irrigated,
459; return yielded by, 459.

Ganges River, basin of, 3; length of,
4; tidal and deltaic, 19, 173, 232;
rise of tide at mouths of, 232; weir
across, for supplying Ganges Canal,
439; weir across at Narora for Lower
Ganges Canal, 440; navigable portion
of, 520; traffic on, diminished by
railway, 520; navigation in delta of,
520.

Garonne Lateral Canal, course and
length of, 349; small traffic on, 485.

Garonne River, ratio of discharge of,
to rainfall, 13; instance of river
emerging into large estuary, 18; ratio
of maximum to minimum discharge
of, 50; rate and cost of dredging with
aquamotrice in, 73; prediction of
floods on, 151; lateral canal in place
of, 349.

Gauge-Tubes, 30–32; principle of, 30;
formula giving velocity of flow from
observations with, 30; description of,
31; method of observing with, 31–
32; limits of use, 35.

Geneva Lake, receives sediment of
Upper Rhone, 16; moderating influence of, on Rhone floods, 17, 169.

Georgetown Incline, 396–397; position and gradient of, 396; description
of caisson conveying barges on, 396;
method of working, 396–397; time
. of transit on, 397.

German Waterways, 505–510; chief
rivers forming principal, 505–506;
navigable lengths and depths of, 505–
506; principal navigable tributaries,
506; canalized rivers, and sizes of
locks, 506–507; canals connecting

rivers, 507; depths of, and sizes of
locks on canals, 507; route of projected canal to connect Berlin with
the Rhine, 507–508; lengths of, 508;
proportionate length of, to area,
508; traffic on, 508–509; increase in
traffic on, 509–510; average journey
and traffic on, compared to those on
Russian and French waterways, 519.

Ghent-Ostend Canal, connecting
Ghent with Bruges and Ostend, 491;
level course of, 492; depth of, and size
of locks on, 493; small traffic on, 494.

Ghent-Terneuzen Ship-Canal, 554–
556; discharge of dredgings for enlargement of—by chain of buckets,
93; by floating tubes, 94–95, 355; by
long shoot, 95;
excavators used in enlarging, 95–
96; dimensions of, 351, 555; forming
short route from Ghent to sea, 493,
555; traffic on, 494; large increase
of traffic on, 495, 556; early works
previous to, 554; commencement of,
554; extension of, 554–555; improvement of, 555; enlargement of, 555;
effect of enlargement of, on traffic,
556.

Gironde River, instance of estuary
with contracted outlet, 247; forming
outlet to river Garonne, 349.

Glasgow and Paisley Canal, depth
of, 351.

Gloucester and Berkeley Ship-
Canal, 551–552; lateral canal to
tidal Severn, 350; dimensions of,
351; route proposed for improved
waterway from Birmingham to sea,
478; object of, 551; description of,
551–552; cost of, 551; proposed
improvement and extension of, 552;
estimated cost of proposed works,
552; advantages of, 552.

Godavery Canals, discharge of, 462;
area irrigated by, 462; cost of, per
acre irrigable, 462; return yielded by,
462; used for navigation, 520; regular
traffic on, 520.

Godavery River, anicut or weir
across, 112, 461; delta of, irrigated
by Godavery Canals, 462.

charge of, at Briare, 50; floods of, greater than those of Seine, 150; dams for mitigating floods of, 155; valley of, protected by embankments, 161, 165–166; periodical inundation of valley of, 166, 171; remedy proposed for injurious inundations of, 166; contrast of estuary of, to tideless outlets, 173; mixed origin of alluvium in estuary of, 233; instance of estuary with contracted outlet, 247, 304; comparison of physical conditions with those of Seine, 304–305; discharge of, 305; tidal rise at mouth of, 305; length of tidal, 305; position of Nantes on, 306; tidal flow of, overpowered by large floods of, 306; encumbrance of upper estuary by islands, 306; accretions resulting from early works in upper estuary, 306–307; training works in upper estuary, 306; results of training works, 307–308; lateral ship-canal for avoiding shallow part of estuary, 308, 350, 552–553; dredging in estuary, 308–309; improved navigable condition of, 309; remarks on training works in estuary, 309–310; lateral canal in place of torrential, 349; communication between lateral canal and Briare Canal across, 360–361.

Loire Ship-Canal, 308, 552–553; to avoid part of Loire estuary, 308, 350; course and length of, 308; dimensions of, 308; improved navigable depth up to Nantes by help of, 309; description of, 552–553.

Long Shoot for discharge of dredgings, 93–94; at Suez Canal, 93, 355; at Panama Canal works, 93–94, 355; in enlargement of Ghent-Terneuzen Canal, 95; used on Baltic Canal, 573.

Lot River, angular weirs on, 110–111; canalization of, 480.

Louisville and Portland Canal, for avoiding falls of Ohio, 527; length of, 527; size of locks on, 527, 534.

Lower Ganges Canal, 439–440; description of, 439–440; area to be irrigated by, 440; aqueduct carrying, across Kali Nadi valley, 440; cost of,

per acre irrigated, 459: return yielded by, 459; used for navigation, 520.

Low-Water Line, of Tidal Rivers, 239–240; lowered by deepening of channel, 239;

lowering of, — instances of, by dredging and training works, 240; indicating increased tidal capacity and improved condition, 240; conventional indication of, on sections, 240; on Thames by removal of old bridges and dredging, 244; by works on Clyde, 255; by dredging on Tyne, 263; by works on Tees, 269; by new outfall on Witham, 286; by training of Welland, 287; by straight cut on Nene, 287–288; by straight cuts and training works on Ouse, 288; by dredging in Ribble, 294, 295, 296–297; in Weser by training and dredging, 302; in Seine by training works, 313.

## M.

Maas River, 222–224; tidal and deltaic, 19; dredgings from, discharged by chain of buckets, 93; delta formerly formed with Rhine, 203; deterioration of northern outlets of, 222; Voorne Canal formed to avoid bar at Brielle mouth of, 222; training of Scheur branch of, and new outlet for, 222; jetties at new outlet of, 222–223; enlarging widths of Scheur branch of, 223; formation of new cut for Scheur branch of, 223; rise of tide at mouth of, 223; supplementary works at new Scheur branch of, 223–224; dredging at outlet channel of, 224; improved navigable depth of, 224; shortened time of transit between Rotterdam and the sea, 224; increased traffic on, resulting from works, 224; cost of works at Scheur branch of, 224; devoid of estuary, 227; rate of enlargement of, from Schiedam to mouth, 324.

Mahanuddee River, delta of, irrigated by Orissa Canal, 462.

Mahmoudieh Canal, to avoid alluvium at Nile outlets, 199; course and length of, 199; cost of, 199.

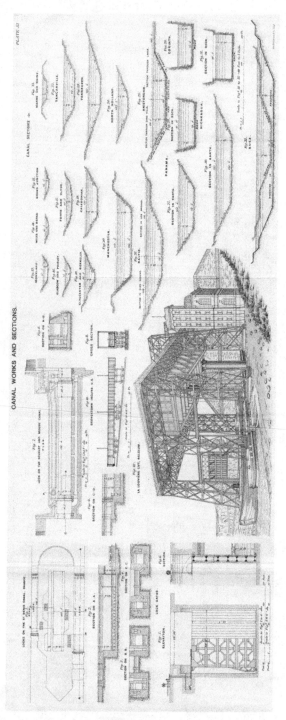

The material originally positioned here is too large for reproduction in this reissue. A PDF can be downloaded from the web address given on page iv of this book, by clicking on 'Resources Available'.

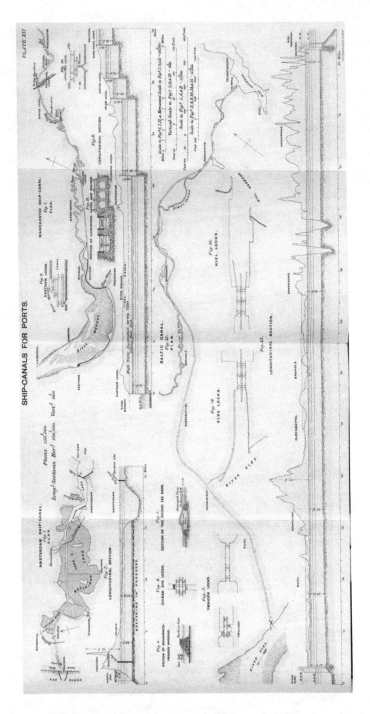

SHIP-CANALS FOR PORTS.

PLATE XIII

The material originally positioned here is too large for reproduction in this reissue. A PDF can be downloaded from the web address given on page iv of this book, by clicking on 'Resources Available'.

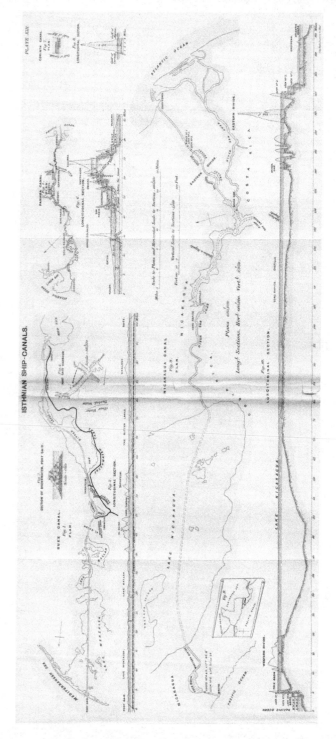

ISTHMIAN SHIP-CANALS.

PLATE XIII.

Printed in the United States
By Bookmasters